CASTING NEW MOLDS

First Steps toward Worker Control in a Mozambique Steel Factory

A conversation with
Peter Sketchley and Frances Moore Lappé

Also by Frances Moore Lappé

Diet for a Small Planet (Ballantine, 1975)
Food First: Beyond the Myth of Scarcity (Ballantine, 1979)
World Hunger: Ten Myths
Mozambique and Tanzania: Asking the Big Questions
What Can We Do? A Land, Food, Hunger Action Guide

To order additional copies of this book:

1 to 6 copies	$2.25
7 to 49 copies	$1.80
50 or more copies	$1.35

Add 10 percent of total for postage and handling. Make checks payable to the Institute for Food and Development Policy, 2588 Mission Street, San Francisco, California 94110 U.S.A.

ISBN 0-935028-08-0

ACKNOWLEDGEMENTS

This book might never have happened without the hospitality of Flo and Jack Westoby who provided just the right setting for our first conversation leading to this book. We thank you.

We are also indebted to friends of ours and of Mozambique who gave generously of their time to offer very useful criticisms of our draft: Nick Allen, Barbara Barnes, Joseph Collins, Steve Goldfield, David Kinley, John Loxley, Brian McCall, John Saxby, John Saul, and Jack Westoby.

We thank Ina Moore, Lisa Kokin, Jennifer Lovejoy and Stephen King for their help in typing our tapes and the book itself. And we appreciate the careful editing, proofing and design work of Jennifer Lovejoy and Doug Basinger.

Finally, we are grateful to Lisa Kokin for the fine art work that brings CIFEL's story to life.

TABLE OF CONTENTS

Introduction 5

Glossary 9

Chronology 11

I. Colonial Inheritance 13

II. Technical Assistance: The role of foreigners 18

III. Leadership: The role of FRELIMO 24

IV. Compromises or Retreats? 35

V. Planning 39

VI. Reflections 49

Conclusion 56

Resources 59

Introduction

Much of the work of the Institute for Food and Development Policy is an effort to step beyond generalizations and into the real life struggles of individuals and groups working to overcome the obstacles to social, economic and political development. This conversation with an Englishman who worked for two and a half years in a steel factory in Mozambique is no exception. Its purpose is to provide a firsthand, personal account of what it is like to live and work in a society newly liberated from centuries of colonialism and setting out to build a new form of social organization, incorporating economic justice and worker participation.

For those of us living in capitalist countries, it is difficult to learn from the day-to-day experiences of people living in countries whose governments' stated aim is creating a socialist society. Reports in the Western mass media focus primarily on policy statements of government officials. Portrayals of daily life tend to be based only on superficial impressions; often only failures and shortcomings are reported. On the other hand, those sympathetic to efforts toward socialism sometimes err in the opposite direction, presenting an idealized picture of progress.

Both approaches can lead to cynicism. Neither approach prepares us to appreciate the complexity of the process of social transformation. Nor do they allow us to gain inspiration from the real achievements of our counterparts. But because Peter Sketchley draws us into the day-to-day ups and downs of the struggles of the workers in this factory, the picture that emerges in this conversation is not an idealized one. Because of the strength of the people we meet as well as the changes--both agonizingly slow and dramatic--we witness through Peter's eyes, we cannot help but also be inspired.

THE AUTHORS

Peter Sketchley was born in Loughborough, Leicestershire, England, in 1947. His father was a sheetmetal worker. Peter left school at 16 to become an engineering apprentice in a large factory. Studying part-time, he specialized in systems analysis for computerized production control systems. During the ten years he worked in England as a systems analyst, Peter was actively engaged in trade union work, serving both as a shop steward within his factory and in a national trade union post.

Peter went to Mozambique in 1977 as a solidarity worker through the auspices of the Mozambique, Angola, Guinea-Bissau Information Center in London (MAGIC). His salary in Mozambique was paid by the Commonwealth Fund.

Born in 1944, Frances Moore Lappé grew up in Fort Worth, Texas. She is the author of Diet for a Small Planet and co-author of Food First: Beyond the Myth of Scarcity (Ballantine Books, revised editions, 1975 and

1979 respectively). In 1975 she and Joseph Collins founded the Institute for Food and Development Policy, based in San Francisco.

Frances is also the co-author of Mozambique and Tanzania: Asking the Big Questions, published by the Institute in 1980. This book probes the broader social, economic and political goals of the Mozambique people and therefore provides useful background to the story told here. Other resources for learning about Mozambique are suggested at the end of this book.

On an investigation for the Institute, Frances visited Mozambique during the summer of 1978, the beginning of Peter's second year at CIFEL. Discussion between Peter and Frances which began then in Mozambique continued through letters. Meeting a year later, Frances and Peter decided to try to structure their dialogue so that it might be of use to others. Casting New Molds is the result.

THE SETTING: MOZAMBIQUE

Mozambique is located in southern Africa, east of Zimbabwe, north of South Africa, and south of Tanzania. After over 300 years of colonial domination, Mozambique achieved independence from the Portuguese in 1975.

After 13 years of armed struggle, liberation was won under the leadership of the Mozambique Liberation Front (FRELIMO). By the time of victory in 1975, FRELIMO had driven the Portuguese out of at least one-quarter of the country and had begun to build in these liberated areas new forms of social and economic organization which involved the peasantry for the first time in self-government. Administering these liberated areas introduced FRELIMO to its post-independence leadership role. In February 1977, just two weeks before Peter Sketchley arrived in Mozambique, FRELIMO officially announced its transformation from a liberation organization to a political party and began the selection of members.

Before independence, about 200,000 Portuguese controlled the economic and political life of the entire country of eleven million. Ninety-three percent of the population were kept illiterate and only a handful of Mozambicans had a university education. At independence, 90 percent of the Portuguese fled the country, taking with them virtually all of the technical and managerial knowledge necessary to run the economy.

The flight of the Portuguese was not the only crisis that compounded the tasks of overcoming the country's profound poverty and dependence: during the first five years of independence heavy destruction was caused by regular ground and air raids by Rhodesian forces attempting to destroy Mozambique's capacity to help the Zimbabwe liberation struggle. By the time of victory in Zimbabwe in 1980, the cumulative cost was approximately $500 million. Moreover, Mozambique has been hit by an unprecedented series of floods and droughts, greatly heightening its dependence on food imports at a time when no foreign exchange is available to pay for them.

THE SETTING: THE FACTORY CIFEL

Peter Sketchley tells of his life at CIFEL, a cast iron and steel foundry and steel rolling mill that is one of the largest of its kind in Africa. Located in Mozambique's capital city, Maputo, CIFEL produces basic agricultural implements, wire used in cattle ranching, and steel used in building.

Before independence, CIFEL produced 44,000 tons a year. After the flight of the Portuguese, production plummeted to just over 6,000 tons in 1976. By 1979, Peter's last year at CIFEL, production had reached 13,000 tons but was limited by the small amount of foreign exchange that could be allocated to steel.

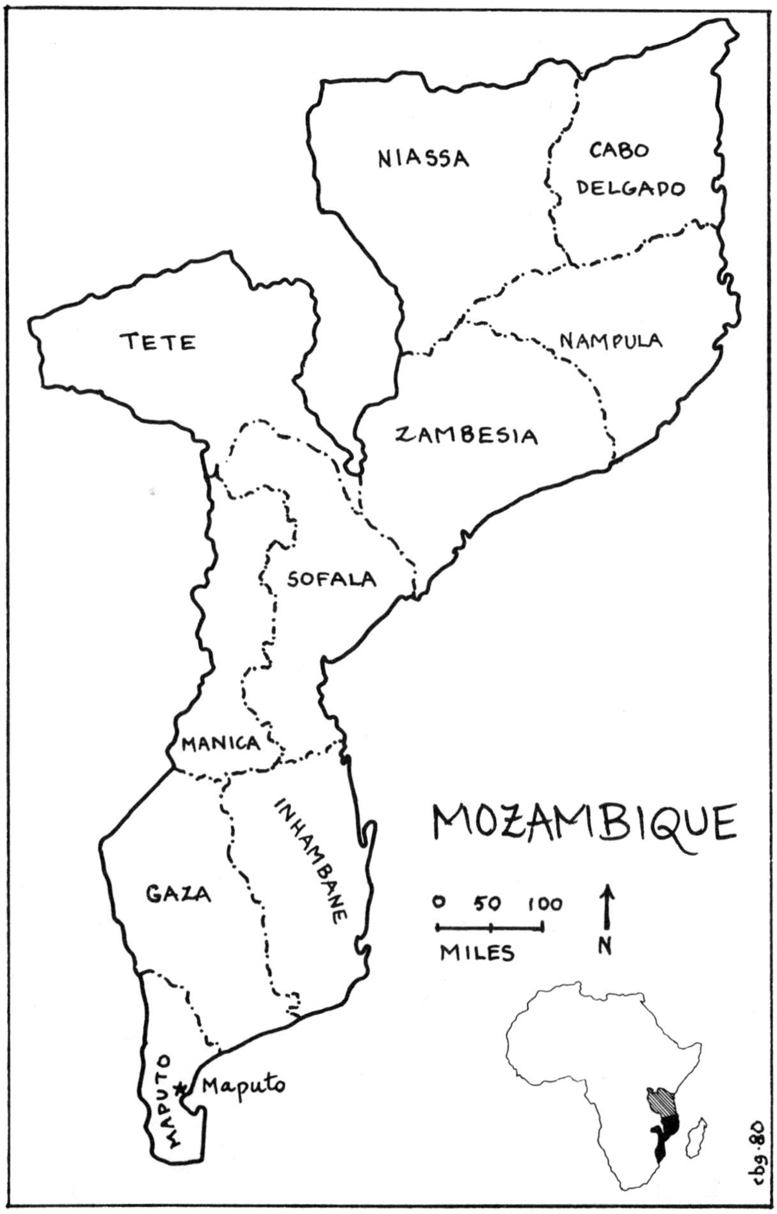

GLOSSARY OF TERMS USED

The factory CIFEL (Companhia Industrial de Fundicao e Laminagem)

CIFEL is one of the largest companies in Mozambique. By African standards it is an enormous plant--probably the biggest rolling mill in Africa outside of South Africa.

It has three sections:

1. The steel rolling mill

Consists of a simple manual mill and also an automated mill. The combined capacity is of the order of 80-100,000 tons per annum. (That is ten times the capacity of a similar mill in Tanzania.) The finished product is steel bars and rods (6-50 mm in diameter and 12 meters long) used to reinforce cement in the construction industry.

The raw material is billets of pure steel (5 inches square and 3 feet long) which are heated to 1000 degrees C and progressively passed through a series of rollers to the required diameter. The billets always were and still are imported from South Africa.

2. The foundry

An electric ore furnace and Bessemer converter supply molten steel and two Cuppola furnaces supply cast iron to the foundry. Objects are cast by pouring molten steel into boxes of compressed sand, filling the impression left by wooden patterns. The total capacity of 3000 per annum is to be increased by 1982 to 8000 tons. The largest single piece ever produced, exceeding 12 tons, was cast in 1978. It well surpassed the pre-independence record of five tons.

The finished product has changed from luxury goods and public service fittings for the Portuguese residents' "concrete cities" to machinery parts, hoes and plows, and essential equipment for sugar and cement industries.

The raw material is local scrap steel and iron but also coke, sundries and equipment from South Africa.

3. The bright wire drawing section

With a capacity of 3-4000 tons, produces bright drawn, annealed and galvanized wire for agricultural, house building and industrial purposes, and also for nails.

Raw material is coils of 6 mm steel from the rolling mill.

Factory Organization*

Dynamizing Groups

Beginning in 1974, dynamizing groups were formed by FRELIMO in each neighborhood, factory and farm. Their tasks were to explain the policies of the government, fight sabotage and mobilize the workers to maintain production in the factories and farms.

Their effectiveness varied enormously--from excellent to very weak and corrupted by opportunists and special police agents.

They were progressively disbanded and replaced by grassroots groups of FRELIMO members, beginning in 1978.

FRELIMO Groups

In February 1977, FRELIMO decided to begin the process of transforming itself from a peasant guerrilla army to a disciplined, leadership party. The selection of FRELIMO members began. Under a system of strict oversight by the people with whom the candidates worked and lived, thousands of workers and peasants were admitted into FRELIMO.

Production Councils

From 1976 on, FRELIMO called for the establishment of Production Councils in each place of work in the country. For the most part selected by the workers themselves, the Production Councils are responsible for the economic, social and cultural aspects of the work place. Subcommittees deal with problems of health and safety, training and discipline. Production Councils can be thought of as embryonic trade unions, although with much wider responsibilities.

Administration

Appointed by the state, it has executive power in the enterprise.

---

*For much more detail on the selection process and responsibilities of the structures outlined here, please consult Mozambique and Tanzania: Asking the Big Questions and A Luta Continua--both listed among the resources at the end of the book.

CHRONOLOGY

1964 FRELIMO (Front for the Liberation of Mozambique) launches liberation war against the 300-year-old colonial control of the Portuguese. (By independence in 1975, approximately one-third of the country had been liberated by FRELIMO.

CIFEL opens with 300 shareholders. Majority of shareholders are the Portuguese Buccellato Brothers.

1968 Rolling mill at CIFEL opens. Major shareholder is Sr. Champalimand, a Portuguese multinational entrepreneur.

1974 April 25. Coup in Portugal overthrows Caetano regime, leading to peace negotiations and independence for Mozambique the next year.

First Dynamizing Group formed at CIFEL.

By end of year exodus of technicians almost complete. Raw materials rapidly depleted, layoff of some workers.

1975 June 25. Complete independence for Mozambique.

First of a series of short-lived administrators appointed at CIFEL.

1976 CIFEL's Production Council forms following President Samora Machel's speech to workers on October 13.

Flight of administrator to Swaziland with "toothache."

1977 March. Peter arrives at CIFEL.

July. Administrator appointed who also owns his own factories.

December. First attempt at "Socialist Planning."

1978 January. FRELIMO Brigade for restructuring the party arrives.

February. Dynamizing Group disbanded.

May. New administration takes full control.

June. Literacy campaign starts, 350 workers start first-level course.

July. Provisional Party branch forms with eight members.

November-December. Second plan made.

1979 May. Planning Department set up and course starts.

August. Peter leaves Mozambique for England.

Production Figures

	1973	1974	1975	1976	1977	1978	(plan) 1979
Rolling mill	44,000	28,000	9,000	6,000	9,000	10,000	12,000
Foundry	2,000	1,700	900	600	300	700	800
Wire drawing	2,500	1,700	950	1,800	1,900	2,400	2,800

Work Force

	Mozambicans	Portuguese
1973	750	100
1977	400	
1978	450	8
1979	600	6

Salary Range

Minimum salary 1974	$27 per month (U.S. dollars)
Maximum manual worker 1979	$450
Majority salary 1979	$120-150
Administrator 1979	$750
Highest foreigner 1979	$1350

I. Colonial Inheritance

FRANCES: Before we get into detail about your experience working at CIFEL, what struck you first when you arrived in Maputo? That was 1977 and free Mozambique was only 18 months old.

PETER: To grasp what it was like, try to fantasize what might happen inside factories like Ford or Bethlehem Steel if the majority of the management, and all of the shareholders, left within a few months for Brazil! And this happened within the context of a breakdown of the police force and disorder.

The workers were suddenly faced with having to keep the factories going themselves. And many Portuguese wrecked everything they could before they left--so factory installations as well as apartments would need major repair before they could be used. Most of the workers who had to face this challenge were illiterate, many couldn't even write their names or speak the national language. Imagine all this, and you start to get some feel of what it was like to be in Mozambique at that moment.

I arrived with my wife and two children the first week in March 1977. It was just at the end of the rainy season, a time of exceptional flooding, which meant that most of the bridges were washed away and the roads were closed.

My first--and enduring--memory was the completely nonviolent and tranquil character of the Mozambican people. In fact, throughout our 2½ years in Mozambique we never heard of a single assault or violence of any kind with the exception of a burglary just before we left. The only thing disturbing the peace of those early days was the clatter of hammers in every corner of the city as the departing Portuguese put together their wooden boxes in which they crated up everything they had. Over 90 percent of the 200,000 Portuguese living in Lourenco Marques, as Maputo was called then, left within the first two years of independence. Many left in 1974 and 1975.

This exodus led to great crises in most parts of the Mozambique economy and life in general. Young Mozambicans, with perhaps only three or four years of education, had to grapple with problems they were completely unequipped to handle.

It was a time of shortages of basic items. The first two months really brought home what a luxury the kind of diet we're all used to really is. The week before we arrived a group of Korean agronomists were shipped back to Korea with acute vitamin C deficiency. On several occasions, for example, no sugar or toilet paper could be found in Maputo, often for weeks at a time. When the supply of imported chicken feed ran out, the market was suddenly flooded with chickens the size of starlings--and with the prospect of no eggs for a long time!

For me it was a time of personal crises, not only the normal crises associated with moving with a family--the problems of education for our two children, housing and feeding us all--but my command of Portuguese was barely adequate. There was also the cultural shock of seeing crippled beggars crawling across the pavement, of daily passing through the shantytown belt which encircled Maputo and seeing naked pot-bellied children and pathetic reed houses standing in stagnant mosquito-ridden water. All of this just as it had been for generation after generation of colonial exploitation.

In one shantytown where my wife, Margaret, worked in a self-help house building project, there were nine watertaps to supply 30,000 people. And yet a few blocks away one could see servants standing all day sprinkling the lawns and gardens of the big homes.

The "developed" part of Maputo, called the "concrete city," is more beautiful than many European cities. Here the Portuguese created for themselved, not simply all the comforts of home, but all the comforts they desired in Portugal but never could have achieved there. Here they lived in a separate town with wide boulevards and tastefully appointed homes.

The few Portuguese who stayed after independence played a wait-and-see game at first. Some suddenly were overwhelmed with the desire to serve the people and stepped in to fill the breach. Some, like the governor of a bank, stepped out again after only a few months with a bag full of money! Similarly, the first administrator of the factory where I worked, after declaring himself before the workers "a true patriotic democrat," left within six months for neighboring Swaziland with a toothache--never to return.

During this change and confusion I felt that the Mozambican people were also playing a wait-and-see game. What will the new FRELIMO government really be like? Will it be *our* government, will it match our expectations? Rural people hungry for land, urban people--now at last with a little bit of money in their pockets--eager for such luxuries as bread and meat to eat in their 10th floor apartments, taken over from the departing Portuguese. That was Maputo in 1977.

FRANCES: To be able to understand your experiences at CIFEL, it would help to know what the factory was like before independence.

PETER: About 750 people worked in the factory. About 100 were the white Portuguese who controlled everything. A few black Africans worked in the accounts and commercial departments. These were *assimilados*--black Africans who, in order to pass to secondary education, had been forced to deny their own culture. They had to go through the most bizarre tests--like proving they could use a knife and fork and didn't sit on the floor and that they ate "civilized" food instead of local food. Some gained minor positions in the Portuguese colonial hierarchy and are still a significant social stratum.

FRANCES: Were the workers organized?

PETER: In the colonial period the workers had no independent organization; certainly there was no trade union, although in the later stages, the Portuguese tried to form the sort of trade unions that Mussolini and Hitler had, structured as part of the state apparatus. But these unions had no rights. they were just to distract the workers from taking any initiative in defense of their own interests.

Throughout the whole colonial period there were no trade union-type struggles in the factory. This is hardly surprising given that the workforce was well infiltrated by PIDE (Portuguese secret police) agents. The workers tell of how the PIDE affected factory life. Workers might all be together talking and in would walk someone known to be a PIDE agent. Immediately everyone would nervously jump up: "I've got to go and do this, I've just got to go and do that." There were other, more subtle examples, where an agent would come and join the factory and within days he'd be saying things like, "Aren't the bosses awful here?" "What do you think of so and so?" This infiltration was a crude attempt to maintain control over the workers.

Anyone who made any gesture of defiance was caught. With a great deal of ceremony, the Portuguese army or police would arrive in force and the workers would be lined up to watch the offender being dragged off never to return to the factory.

Becuase of this absolute control, any sense of unity and struggle among Mozambican workers--virtually all of them of the first generation in from the villages--was incredibly deformed, compared with any of the European working classes. There was no basic experience of even how to organize an agenda or run branch meetings or organize membership and monthly dues accounts, how to set priorities, delegate responsibilities--things which workers in many other countries take for granted.

Also there were psychological barriers to overcome. In Britain we are starting to think about how a woman develops an inferior self-image. But what about the black worker? Beaten, insulted, fined for nothing, constantly under observation, perhaps by the man he is working with. It's interesting to note that the mechanics in the factory who had worked in the South African mines would, when struggling with a rusted nut, often yell the only three words of English they know--"You black bastard"--at the nut.

FRANCES: All of this would have to be shaken off before a worker could express an opinion or take an initiative toward a new way of relating in the factory.

PETER: And how easily he could fold, I found, if a white man postured in front of him like they did in the old days.

FRANCES: CIFEL is one of the biggest industrial plants in the country. How did the new government see the role of CIFEL in dealing with the multiple crises facing the economy?

PETER: CIFEL is important to the political as well as economic base that FRELIMO is trying to create. The concept of building a sense of unity and mutual dependence between peasants (90 percent of the population) and workers is very consciously promoted in Mozambique.

CIFEL demonstrates such a worker-peasant alliance in its very products. The factory, for example, makes plowshares for animal traction plows and the traditional big African hoe. It produces giant rollers used in sugar refineries, and the equipment needed to prepare cement. The wire drawing section of the factory makes galvanized wire for cattle fencing and for strapping up bundles of cotton--for a thousand and one different agricultural uses. The steel bars from the factory are used in the construction of bridges and houses and hospitals. So, in fact, and this is often underlined in political meetings in the factory, the fruits of the production of the workers go directly to assist the peasants in the countryside who in turn must feed the cities.

When Marcelino dos Santos, the planning minister, visited the factory to launch the literacy campaign, he asked in a talk to the workers: "What's the difference between the products now and the products as they were during the colonial period?" The difference is in how the steel is being used, they answered. Not to build luxury apartments, luxury hotels, night clubs; not to build jails run by the Portuguese colonialists. Now it is being used for construction of hospitals, dams and roads. Dams and bridges are being destroyed every week by the Rhodesians. And the government has pledged itself to repair them immediately--ten times a year if necessary! That's one fundamental difference--the difference in the purpose of production now as opposed to what it was.

In production and planning meetings, workers are reminded of what it is they are actually doing--their product, and its purpose. It's always explained within the context of a worker-peasant alliance--"The wire is needed out there." "There is a complete lack of nails in the countryside." "Animal traction plowshares are fundamental in the most remote areas."

FRANCES: How did the workers respond? Did these explanations have meaning to the workers? Did they see their work any differently than under the Portuguese?

PETER: As I have said, most workers are the first generation of peasants from the countryside. When they go back to the villages during their month's holiday when the factory closes, they live the life of the peasants. So they know full well that these shortages are not going to diminish until they resolve the problems in the factory. And they discuss this. I don't mean a kind of superficial repeating of slogans from the party. These people sincerely believe in the worth of the contribution they are making. A different attitude is developing. No, not a completely different attitude. I don't want to idealize it, but in the factory there is a growing awareness among workers of *why* they work, of the contribution they make.

For example, in July 1979 there was an appeal made for workers to assist for the second year in the vital rice harvest in the Limpopo valley. Immediately workers volunteered to go. Participating meant spending a weekend without pay, sleeping rough and perhaps without too much food. It meant a long journey for which they would have to get up at four o'clock in the morning to catch the truck. Collecting rice is hard work too, not romantic at all. And that particular weekend it rained like mad and, by Mozambican standards, it was very cold. But when they came back, I didn't hear anyone say "Oh, Christ, it was cold," or "the food was awful." If you asked them why they went, they said, "Well, because it was necessary, because there are queues for rice and if the people don't solve these problems themselves, the government is not going to be able to solve them."

All that is not to say that all the workers are paragons of social consciousness all the time. That would be a total distortion. There are many examples one can cite at the factory of a section finishing its particular task and sitting down rather than jumping in to help the next section. Nor would I claim that the social consciousness I've seen in CIFEL is replicated in every other factory in Mozambique. Certainly not. This unevenness is unavoidable. Part of the party's role is to stimulate the changing of attitudes toward work, to explain to the workers their new responsibilities and how they serve the reconstruction of the country. So changes in attitude vary in part according to the quality of the people in the party structures in the places of work and in the neighborhoods. It's inevitably been uneven.

II. Technical Assistance: The role of foreigners

FRANCES: Given what you have said, Peter, about the oppression of the workforce under the Portuguese, it is difficult for me to imagine how the workers kept the plant operating after the Portuguese left.

PETER: It is impressive that the workers didn't let the factory close. As I talk about all the factory's problems--all the mistakes, too--we should never lose sight of the fact that, given the circumstances, just keeping the factory going was a great victory!

But let me first explain what the factory is. The rolling mill consists of a furnace which heats huge bars of steel to 1000 degrees C and then passes them, as white hot billets, through a series of rollers (very similar to the rollers used during our childhood to squeeze water out of clothes). They become progressively thinner and longer until at the end they are 6 mm to 50 mm thick and many meters long. These bars are used by the construction industry to reinforce concrete.

There are two mills in the factory. One is manual. Between each set of rollers, the workers--with a pair of long steel tongs--grip the end of the white hot bar and pass it manually into the next set of rollers. This is a rather dangerous operation. And when the temperature outside is over 100 degrees F you can imagine what it's like swinging a white hot bar of steel!

The other rolling mill was automated in 1974, ironically just before the factory was abandoned by the Portuguese. Drive controllers determine roller velocity very precisely, because, obviously, as a bar becomes thinner it's got to travel faster, since the same amount of steel is going through. The rollers have to be controlled with small microcircuits.

FRANCES: I can imagine the difficulty the workers had mastering this type of technology.

PETER: Remember that in 1974, 85 percent of the workforce of CIFEL could not read or write. In June 1978, 350 workers started first-level literacy courses. A theme of the campaign was that every worker had the responsibility to participate to master the technology and thus prepare himself to be the leading force controlling the factory. The slogan for literacy was "Participate in the literacy campaign to dominate science and technology and win the battle for production." (Another slogan, to digress from the point, was "Participate in the literacy campaign to harmonize and equalize the relationship between men and women.") Even after the literacy campaign began, the majority of the workers were at the "a cat, a dog" level of literacy. Repairing an electronic circuit was out of the question. The universities that existed under the Portuguese produced some qualified technicians but the vast majority were, of course, Portuguese, and most of them left. There were fewer than eight Mozambicans in the university in the colonial days.

Technicians, or even people with basic electrical knowledge who could be trained to understand how the factory works, just didn't exist in Mozambique in the numbers required to deal with the present level of technology and uphold the present level of production, much less advance. So help from foreign technicians has been a necessity. It will be for a long time.

One production disaster might demonstrate for you the problems workers face with technology they were denied knowledge of for so long. In the foundry, castings are made by taking a wooden model and then compressing it into a steel box filled with sand. The model is taken out and the space in the sand is filled with molten steel. The sand itself has to be especially prepared with additives--some that bind it together and others that keep it from burning.

We once went through an absolutely disastrous month at the foundry in which the quality of plowshares plummeted. Most of them just had to be scrapped.

An investigation revealed that the section preparing the sand to make the molds into which the molten steel would be poured had to add one measure of white powder to bind the sand and one measure of another kind of white powder to prevent the molten steel from burning the sand.

When one powder ran out (through poor stock control), the worker simply doubled the quantity of the other powder. As a result, when the steel was poured, the sand collapsed.

When confronted by the Workers Council and surrounded by a mountain of scrap metal, the worker concerned turned to us movingly and said, "Don't be angry, comrades, remember in the colonial days, the fascists never explained to us what these were for. As far as I am concerned they are just two white powders."

So then someone had to identify the powder; find its English name and the companies in South Africa that had supplied it; calculate the usage per month; write a telex in English (no one spoke English); get approval from the Ministry of Industry; and get a letter of credit through the bank (where the order would sit in a mountain of requests for foreign exchange which was just nonexistent). This process could, and often did, take months to complete --and during that time production was halted or switched to something else.

There are innumerable similar problems faced daily by the workers. For example, the moisture content of the sand is critical in steel production. If the steel is poured into excessively wet sand, steam is produced and bubbles form on the surface of the finished product.

To improve the quality of the products, an investigation was launched to determine how the moisture is controlled. The sand laboratory still functioned, it was learned. Every day samples were dutifully taken and analyzed for various characteristics, including moisture content. The results were recorded and then filed.

During the investigation the workers said, "Oh, well, that's very good. We're controlling the moisture." But they weren't controlling the moisture; they were only *measuring* it, which is a far different thing. As it turned out, no one even knew what the actual moisture level should be, whether two percent or ten percent or whatever. So, while the analyses were scrupulously performed, they were totally useless! Just as in the case of the additive, the workers had no idea why they were performing a certain function and were unable to take corrective action when a problem arose.

Such is the level of understanding the Portuguese colonialists passed on because of their sense of racial superiority and because it was in their interest not to promote an understanding, but rather a mystification, of the production process in the minds of the Africans.

An equally common problem can be seen in the example of the thermometer which is used in the electric arc furnace to measure the temperature of the steel before it is poured. The temperature is critical and directly affects the quality of the product. If the steel is poured too cold, it doesn't fill the mold correctly and causes premature solidification. If poured too hot, it burns the surface of the sand and crystallizes it, resulting in poor surface texture.

The thermometer broke down in 1976. Since then, the temperature has been controlled on the basis of how it appears to the eye of the furnaceman. This eye-balling method is inadequate but the obvious solutions--repairing or replacing the defective equipment--have been impossible.

Once an attempt was made to contract with a South African engineer to repair all the thermometers in the factory. He accepted a huge payment, failed to fix the thermometers, and was never seen again! This is not uncommon and, of course, creates in the workers much hostility and cynicism toward such people.

Looking the other day at giant castings made 150 years ago in Sheffield, England, I was thinking that they couldn't have had transistorized pen recorders in those days, or hygrometers to measure moisture. What they did have were much more simple but very adequate techniques. For example, they would pour a little molten metal into a fire clay pot and count the number of seconds until an oxide film appeared on the surface. Experience would tell them how many seconds corresponded to the temperature they needed for a particular piece. Similarly, a test ball of sand was rolled in the hands and then broken in half. The break should be clean and without crumbs. The ball should also be able to take the print of a thumbnail. What furnacemen of the industrial revolution lacked in electronics they made up for with a long and rich hundreds-of-years-old craft culture.

The difference in Mozambique is that none of the workers at CIFEL had any historic roots in the iron and steel smelting which was going on in what is now Zimbabwe and was being shipped through the port of Sofala before Portuguese explorer Vasco da Gama was born. Those cultural roots were broken during the terrible years of colonialism. History now demands that overnight these workers must close that gap.

FRANCES: You mentioned that the government has been forced to look to foreigners for help because few Mozambicans have the needed skills. But which foreigners? And what has been their impact?

PETER: The vacuum left by the departure of Portuguese technicians has to some extent been filled--on a stop-gap basis--by the government's contracts with foreigners. Agreements have been signed with sympathetic countries, particularly of the Eastern bloc. Also, contracts are given to Portuguese who were previously filling the positions, or to Mozambicans who opted for Portuguese citizenship. There has also been a mass recruitment of foreign experts called *cooperantes*--best translated as "cooperators." They are selected through solidarity committees, in, for example, Britain, Holland and Sweden.

It's important that Mozambique is determined not to become beholden to any block or superpower state exclusively. Its stated policy is to seek assistance from a diversity of sources. In this way Mozambique hopes not to be caught in a desperate situation, as China was in the fifties when Soviet technicians left, or as Vietnam was when China pulled out hers.

The role played by the foreign community is very mixed. One example is the Portuguese head of the mechanical maintenance section of CIFEL. This man is a complete individualist, utterly arrogant and paternalistic. For instance, when a piece of plate steel is ready to be cut with a flame cutter, the workers turn to him to mark it out and he does it. He doesn't show them how he does it, nor encourage them in any way. He performs a series of key tasks like this which are perceived as being his and his alone, and he does nothing to alter that perception.

Once I telephoned a clerk in the maintenance section to ask for the serial number of a particular machine. All it involved was reading a number off a plate, but the clerk said, "Oh, you'll have to wait for Señhor Martinez. He knows. He is the one that understands these things."

This one illustration tells so much, not just about this particular man, how he worked and the effect he had upon the workers, but about the whole colonialist mentality and how it could erode and finally destroy any perception the African workers might have of themselves as capable of taking initiative and accepting responsibility. His authoritarian and racist approach, full of bantering references to the workers having come from the jungle and being therefore incapable of understanding how the modern world works, only served to reinforce the already serious inferiority complex of the Africans.

My experiences in Mozambique leave me with a rather hard and critical view of the Portuguese who are now working there. Of course there are notable exceptions--young, progressive Portuguese who came to Mozambique after independence. But they are an exception and perhaps that is not surprising. Living under Salazar/Caetano fascism with its repressive, bureaucratic and racist arrogance means that the Portuguese suffer from a deformation similar to that of many English toward the Irish. Also, there are many opportunists among the Portuguese--unemployed technicians from Lisbon with the ink barely dry on their Portuguese Communist party cards!

FRANCES: What about the influence of non-Portuguese foreign advisors?

PETER: It is very mixed. From time to time delegations from places like the Soviet Union, East Germany, Romania or Bulgaria would come round to the factory and were shown everything. They were told about the problems of illiteracy and the technical problems with the machines. They were shown the quality of the products, the production rates and what have you. And very often, at the end of the tour, they would sit down with the representatives of the Administration Commission, Workers' Production Council and the party unit, and they would be asked, "Well, you've seen the factory and all our problems, which we have made no attempt to disguise. As specialists in this field, what do you advise? What priorities, what measures would best help us achieve the production goals we have set?"

Typically they would reply: "Well, you must immediately reorganize and mechanize the plant, particularly the foundry. There's far too large a labor force and it can only be overcome by implementing automated carousel-type equipment like this, for example." And at that point out of the briefcase would come a glossy catalog of the equipment produced in their country, which they would then proceed to try to sell.

"But," the factory representatives would say, "how can this be useful to us in Mozambique? Here we have an enormous fund of unemployed and underemployed people. Here we have no foreign currency at all at our disposal."

And the standard answer was, "Well, unemployment is a problem at the level of the factory. They are two separate problems."

And that, I submit, is perhaps not the most helpful way of looking at the development possibilities for Mozambique!

Of course, there is also a tendency among Mozambicans, and most in the third world, to be fascinated with the very latest and to feel that the latest, the newest, is the best and that any other kind of technology is settling for second best, a sort of poor man's solution to the problem. Unfortunately, a consideration of alternate, more appropriate technologies doesn't seem to have been developed in many of the Eastern bloc countries. But then, and quite naturally, the multinational companies are not going to offer appropriate technology solutions, are they? And even United Nations aid agencies with their highly paid experts seem insensitive to the problem.

At this point I want to underline one thing and that is that much of the criticism of the help that's been given by the Russians, East Germans, Bulgarians, etc., is made by people who are basically opposed to any measures FRELIMO wants to implement. They are part of an orchestrated campaign to discredit the gains of FRELIMO and the real contributions made by some very dedicated Cubans, East Germans, and so on. I think it's important that one disassociate oneself and one's criticisms from the criticisms made by those who want FRELIMO to fail.

There are campaigns of whispers and rumors that continuously circulate --for instance, about East German trucks being too high off the ground so they will leave the road and kill the drivers, that doctors from such and such country are not to be trusted, etc.

From my experience, technically the East Germans and Soviet bloc contributions are usually good, often excellent. But for my taste, the implementation tends to be bureaucratic and formal and lacking any sensitivity to what the Mozambicans call "people's power."

On the other hand, I have seen Cubans voluntarily harvesting rice (and setting a cracking pace for the rest of us) and having a lot of fun at the same time. They are paid a few dollars a week pocket money and are all genuine volunteers.

Nordic aid, particularly that from Sweden, is worthy of special note. Substantial financial support was given to FRELIMO by the Swedish social democratic government even during the liberation war. In recent years many agricultural projects, coordinated efficiently by the Swedish government aid agency, have been refreshingly sensitive to Mozambican aims and needs. Moreover, Nordic aid in general has tended to be more in grants, whereas a large proportion of aid emanating from Eastern Europe is in long-term credits.

This supportive role of the Swedish government is hardly pure altruism. It has also been a boon to the entry of Swedish capitalists! A great joke among foreigners in Maputo was the suggestion that Volvo set up a car assembly plant. The cars would have arrived in five pieces and they would have cost more than the assembled ones!

There is also a small band of volunteers from the U.S.--fine people, working all hours as doctors, nurses, teachers and technicians. You met some of them, Frankie. Wouldn't you agree that any country could be proud of them? I have seen many a Mozambican scratch his head and try to reconcile these good Americans with what he knows of United States imperialism.

III. Leadership: The role of FRELIMO

FRANCES: You're a socialist, and for many years have studied the theory of societies in transition, as Mozambique is now. Do you feel, as a result of living the daily experiences you have been describing, that you've come to modify your view at all?

PETER: Oh, yes. I can feel my skin creep as I think of some of the naive preconceptions I took with me to Mozambique. I really believed that socialism was just a question of clearing out the reactionary, repressive regime, and that somehow society would be spontaneously transformed by the people, like some of those super-optimistic films of post-revolutionary China. Well, the reality hits me like a cold shower. Mozambique was certainly not like that. Enthusiasm or, more importantly, the confidence of people to stand on their own two feet, to straighten their backs and raise their heads and say, "We are now responsible for doing this ourselves, and we can do it"--that is the product of a slow and tedious process of consciousness-raising facilitated by the party FRELIMO.

I learned that nationalizing basic industries didn't result in fundamental transformation overnight. (It's possible for a factory to pass into common ownership, but then to continue to operate in exactly the same way it did before.) Nor did the taking of state power by FRELIMO result in an overnight transformation of society. But both were essential, I believe, in that they created the conditions for the possibilities of democratizing the factory and the whole of the state apparatus. But nationalization should not be confused with that transformation or even be seen (as it sometimes is in Britain) as a valid end in itself.

If you talk to mineworkers and steelworkers in Britain who remember the great postwar nationalization in 1947, they'll say that the biggest and most bitter disappointment for many of those who'd struggled for a new society was to find that the day after the nationalization they returned to the mines to find the same managers doing the same things as the day before! The power relationships remained the same. The nationalized sector of the British economy is just a different form of capitalism in which capital shareholders have disappeared but everything else continues.

And this was just the situation encountered during my first year at the factory. In July 1977, the Ministry of Industry appointed an administrator with executive power over everything in the factory. He had been a member of the city council of Lourenco Marques and still owned a couple of small factories. To his credit he made a dramatic flash-in-the-pan improvement in production. A phone call to the bank unlocked a few channels and released the long-awaited letter of credit. Soon the much-needed steel started rolling out the door. Solely in terms of levels of production, things were at last looking up at CIFEL. But beyond that the relationships in the workplace were taking a turn for the worse.

The administrator was busy consolidating a pyramid structure with him as the apex and the six Portuguese technicians as department heads at the base. The Production Council was constantly being reminded that it was first things first and that they should be playing a policing role against indiscipline, lateness and absenteeism. (Of course, indiscipline *was* a problem. Under the Portuguese, people were forced to work and were subjected to iron discipline--fines, beatings and instant dismissals. Avoidance of work was an act of defiance against injustice. Under these conditions, no internalization of work discipline could take place. So, when the iron hand was lifted at independence, the short-term effect was widespread indiscipline, absenteeism and drunkenness.

But whatever sparks of initiative were developing among workers were fast being extinguished. And perhaps the workers felt relief. The reins of power were passing firmly into the hands of the administrator and his capitalist-style structure.

An unholy alliance existed between him and the secretary of the dynamizing group. As I learned shortly before I left, the administrator also persuaded the secretary of the Production Council to visit the factories he personally owned, (as if sent by the provincial headquarters of the Production Councils) to lecture *his* workers about the virtues of discipline and sacrifice.

Thus, the gaining of state power and the nationalizations only created the requisite *conditions*, not the transformation of power relationships. It was then up to FRELIMO to begin those social changes. To do so meant reorganizing itself as quickly as possible from a broadly based peasant guerilla movement into a tightly disciplined party that could provide the needed leadership.

FRANCES: I think the idea of a one-party system frightens many people. Perhaps it's because we can't separate out the idea of the "party" from the corrupt and alienating models of authority which we ourselves have experienced and read about. Also it's in part because of the penetration of the Big Brother-Cold War stereotype.

PETER: Yes, and all this can block our ability to imagine the party as anything but necessarily repressive. But please understand that I'm not suggesting we throw out one stereotype and replace it with its mirror image. It would be too easy for me to paint a rosy picture of FRELIMO. Instead, let me draw on my experience in the factory.

The first embryonic form of political organization to appear in the factory was called the "dynamizing group." The history of the dynamizing group's role in the factory is very important. But first let me explain a bit about the dynamizing groups in general. Immediately after the coup in Portugal in 1974, the first task of FRELIMO was to try to mobilize the people toward immediate and specific objectives. Initially the dynamizing groups were organized by FRELIMO as they developed, then to combat economic sabotage, and finally to mobilize cooperation to overcome the economic crisis precipitated by the exodus of the Portuguese. They were formed in all the major factories, in the farms, communities around the cities and in the villages. In the villages, they were fundamental in dismantling oppressive aspects of the old tribal structures, including the old chieftain system. Some of the constructive functions of the tribal structures passed to the dynamizing groups.

The formation of the dynamizing groups was inevitably haphazard because the party simply was not organized yet to structure it better. Remember, the party began its own organization from a liberation force to a political party only two weeks before I arrived. So many of the most opportunistic elements--even some who had collaborated with the Portuguese and the PIDE--got into the dynamizing groups. Often only the people with experience in executive authority put themselves forward and many of them were egotistic and demagogic.

One was the secretary of the six-man executive of the dynamizing group at CIFEL. He was a young, bright assimilated Mozambican who worked in the laboratory. To give you some idea of his demagogic manner, he would drag a worker to his feet during mass meetings and say "*You*--what does socialist emulation mean? You mean you don't know? Don't you read about FRELIMO? Aren't you interested?"

After he had repeated this a few times he would say, "OK, then *I* will have to tell you. It means produce more and better in less time," or some other glib slogan.

One night I gave him a lift home and, as he waved to the workers, he said, "What a pity there isn't a hole in the roof so I could stand up and all of the workers could see their President." He was young. He was unformed politically and deformed by the process of assimilation. But at 19, could you have led a steel mill? Nevertheless, he was *the* political leadership at that time and hated by the majority of the workers. The understandable effect of his approach was to discourage the workers from participating in the mass meetings.

At that early stage the dynamizing group stressed punctuality and discipline. As I have said, there existed a rather opportunistic alliance between the secretary of the dynamizing group and the old capitalist-style administrator. The secretary of the dynamizing group would launch into great attacks on the workers for their lack of political consciousness and discipline. But everyone in the room knew that in a list ranking the worst offenders for indiscipline and absenteeism, the ten members of the dynamizing group would be in the first twenty places! It was sheer hypocracy.

FRANCES: The dynamizing group you've described fits very well a stereotyped image of the "party"--the demagogic bully-boy trying to use threats and slogans to force people to participate.

PETER: And its effect is not difficult to imagine--disaffection, demoralization and apathy.

But in late 1977 the national campaign to restructure the party got underway. Things began to change. A team from FRELIMO arrived at the factory in early 1978. They were intense young people with clipboards taking notes, gathering opinions.

Within a week they declared the dynamizing group to be a spent force without the support of the workers and closed it down. But in doing this, the FRELIMO team also stressed to the workers the need for unity. Either the members of the defunct dynamizing group who abused their power would come to FRELIMO or they would become disaffected, opposing the party's goals. So the challange now was for all the workers to try to recuperate these people and incorporate them more and more into the political life of the factory without hostility or recrimination.

Then the FRELIMO team called a mass meeting in which all the workers were invited to air all the problems of the factory. Nothing was too big or too small to be raised. Complaints were made about the occasional failure to distribute the daily loaf of bread to which each worker was entitled, lack of boots and safety equipment, problems of essential spare parts, etc.

Then at the end of the meeting, the FRELIMO leader again asked for any other contributions. Nothing is too big or too small, he said again.

Finally an elderly worker whose job was sorting the scrap in the scrap yard came forward and stood at the front of all of the workers in his bare feet and endlessly patched work clothes. He stood there for a moment, with his eyes to the floor.

Suddenly he cleared his throat and in a loud voice cracking with emotion he said: "Imperialism is still here in CIFEL." He went on to denounce the administrator for having nothing but contempt for the workers and behaving just as the others had done before.

He sat down to a roar from every throat in the room. Now everyone wanted to say his piece, often repeating the same stories and attacking the other Portuguese in the factory as well. It was as if this ordinary barefoot worker had pulled the cork out of the bottle and all of the frustration and anger came pouring out.

Every word was duly recorded and the FRELIMO representatives promised that all would be reported to the national headquarters of the party and from there to the ministry of Industry. Within a few weeks the administrator was removed and replaced with two young black Mozambicans, one with executive powers, the other his assistant.

But the main purpose of the team from FRELIMO was to help form a provisional unit of FRELIMO in the factory. Six of the twelve workers proposed for provisional party membership were accepted by the FRELIMO team, but only after every aspect of their lives was discussed critically by the workers, and only after the FRELIMO team had visited the neighborhoods where the candidates lived in order to know whether they were respected and trusted by the people there. The person selected as chairman of the new party unit was a jettler in the factory. (His job was smoothing out rough castings with grindstone.)

With the old administrator gone, with two new Mozambican administrators and with the new party unit in action, life at CIFEL at last underwent a profound change. In this phase it was a pleasure to work at CIFEL.

The administrators and the new party unit visited every section in turn, sounding out problems, making criticism and calling for self-criticism to solve the problems. Where one section had a criticism of another, a joint meeting would be held--like the very heated one between the rolling mill workers and electricians. After letting the shouting and recriminations run their course, the administrator would quietly draw it all together and suggest new procedures and guidelines. After agreement was reached, we would all sing a FRELIMO song and one about solidarity with Zimbabwe.

The whole approach of this newly formed FRELIMO group within the factory was quite different from that of the dynamizing group. In mobilizing and organizing the workers, the heavy, demagogic line was dropped and replaced by a whole series of measures which transformed life within the factory.

For example, six popular culture groups, African traditional dance groups with drums and sticks and animal clad dancers, were formed. They practiced every Friday afternoon right outside my office. Popular culture was part of a national campaign. It was an attempt to break with the colonial mentality and rehabilitate pride in the cultural heritage of the Mozambican people. It made a statement: this is our culture and we're proud of it. These very lively groups performed at the beginning of all meetings and whenever visitors came to the factory.

Secondly, the social center of the factory, previously reserved just for the white Portuguese, was reorganized. At each lunch break, tea was provided for all the workers, who could now sit together at tables. Discussions and socializing at lunch time created a much greater sense of unity and social harmony within the factory. Also, each day the workers formed a circle and the main points of the day's newspaper were read out loud. Each worker was issued a new boiler suit (the first that they'd ever had), safety equipment and boots. On Mozambican Heroes Day, which happened to be the 10th anniversary of the death of Eduardo Mondlane, the first president of FRELIMO, a festival was organized inside the foundry, two buffalo were killed, everyone had plenty of food and beer and the dance group performed.

Well, these are some big and small examples of the efforts of the newly formed party group to undo some of the

LK

disaffection and antagonism which the dynamizing group had generated. It was altogether a much more human approach. The administrators, for example, joined in the weekly production meetings in each section, often sitting on the floor in the circle of workers after joking and teasing with them.

It was during this period that the biggest-ever piece in the history of CIFEL was cast: a 12 ton cast-iron sugar cane crusher. All the Portuguese just laughed at the idea when the Minister swept into the factory on a flying visit to say that there was no money to buy sugar cane crushers and, if we couldn't produce sugar, we couldn't export it and things would be worse for everyone. He made it plain that he was not asking *if* it could be done but, with that infuriating FRELIMO optimism, wanted to know only the date when the first one would be ready!

Teams were formed to study the problems. The furnacemen designed a huge new ladle to carry the molten metal and a reusable molding box was made in a nearby shop.

The cane crusher was cast on the day promised and was perfect. The Minister was delighted and wanted to know how soon they could have 350!

Not everything ran smoothly though--a giant breakdown in communication between the administration and the newly formed state import/export enterprise meant that when we telephoned three months after placing the order to ask the South African mill where our steel was, it turned out that no order had been placed with them.

FRANCES: Can we go back for a moment, Peter? Especially in contrast to the new administrator, that first administrator sounds awful. Why on earth did FRELIMO appoint him in the first place?

PETER: He was appointed by the Ministry of Industry in 1977 at a time when, for reasons we have discussed the factory was virtually paralyzed and not meeting even the low level of demand for steel needed for bridge repairs etc. The 24-person Production Council, which had complete control at that time, was overwhelmed with the task. Stock records had fallen into disuse with resulting raw material shortages. Motors were burning out due to lack of maintenance.

FRELIMO's position since the liberation has been never to prejudge a person based on the superficiality of his race or what he did before. Each person is given a chance to prove his stated loyalty in practice and the party intervenes to try to win him fully and develop his consciousness further. The administrator was not a sinister or evilly-intentioned person but he was cast in a particular mold of the colonial days. He was not able (or in his heart didn't want) to break with all that.

A fascinating question to ask is why the second administrator--who came from an *assimilado* background--*was* capable of transforming himself. This I pondered often looking at many such dedicated young people who were struggling within themselves. On the one hand, they had to deal with an inherited penchant for the good life of individual self-gratification. On the other (as in the case of the second administrator, they truly empathized with the struggle of the ordinary, poor worker. This internal conflict within those wanting to change revealed itself painfully in high incidences of stress-based psychosomatic illnesses like ulcers and chronic headaches.

FRANCES: Sometimes I fear that because we, correctly I think, stress the *social* transformations we see as necessary (the redistribution of economic and political power), it is easy to forget that the significance of any social transformation ultimately hinges on whether individuals can change.

PETER: Yes, I agree, and if we understand this, then we can fully appreciate the contribution, say, of the feminist movement. I should also add that I was most certainly not immune from this traumatic self conflict. Even as a foreigner and therefore not expected to have the same "revolutionary discipline and self-sacrifice," as they called it, I felt very uneasy about skipping around the bay in my little ski boat or eating out too often at restaurants. Yes, and being bugged about having to sign the late book if I overslept. And sometimes I would lose my temper if work which was desperately late would be nonchalantly dismissed with the inevitable promise of being ready "tomorrow."

At times it was just too easy to get out of the heat and hide behind my more comfortable and less demanding role as foreign advisor.

FRANCES: Returning to the important changes that began to occur in the factory, could you explain more fully why the dynamizing group turned out to be so bad and the FRELIMO group, apparently, so good?

PETER: Well, the difference is, of course, the process of selection and control. Remember, the dynamizing group was formed before FRELIMO was organized as a party to oversee the selection of people for positions of power and responsibility. As a result, the process was wide open for the most opportunistic elements--the former PIDE police, the people who had collaborated with the PIDE agents in the Portuguese days--to put themselves forward. Some became members of the dynamizing group in the factory--basically people who weren't preoccupied with the good of the people or the process of national reconstruction. They really cared about what they could get for themselves or being respected, being at the center of things, having power.

FRANCES: But why did the workers allow people like that to become their leaders?

PETER: I once asked Jordao, who worked in the sales department and was one of the closest friends I made at CIFEL, how on earth it was that young Sengo could have been accepted as the dynamizing group secretary. I remember his reply: "But companheiro Peter, this was our very first experience of any kind of election. No one explained what qualities of the person we should be looking for. And Sengo is one of the best educated amongst us. When he said he would like to do it, we were all eager to accept him. Aren't we entitled to at least one mistake? But seriously, the problem was that in those days, we had no way to criticize or to replace him."

Shortly after that conversation, a mass meeting was called to put before the workers a member of the Production Council accused of theft of funds. We were all reminded by a very angry chairman of the party unit that *no one* was above the criticism of the workers, and that included him and, if necessary, the administrator as well! (This brought a whispered gasp from the meeting.) We were constantly taught that it was a privilege to serve the people and unthinkable to abuse that privilege.

To return to your earlier question about why the FRELIMO group was so much better than the dynamizing group, remember that the process of selecting members was very different. Recall that, at the time of restructuring the party, a team from FRELIMO came into the factory to help form a provisional unit of FRELIMO. The initial selection of 12 names was made by the FRELIMO team after it went through every department and asked the ordinary workers on the shop floor: "What do you think about so-and-so? Who do you think would make a good member?" Workers were also invited to put themselves forward and to nominate their fellow workers. That's how these 12 names were all drawn together.

These nominations were announced and then discussed with the workers. Based on these discussions, the 12 were cut to eight. Then each of these prospective candidates had to stand before a whole series of meetings, with the section and department in which he worked and with the whole factory. Every aspect of their behavior was discussed and criticized. The team even visited the party structures in each of the neighboorhoods where these candidates lived to inquire about their lives--what their relationships with their families were and how they contributed to political and social development in the neighborhood.

FRELIMO emphasized many times that to become a member didn't mean one had to be some kind of intellectual or technical expert. It did mean that the person had to be someone others trusted. The candidates also had to demonstrate that they were exemplary workers.

Of the eight selected initially, all but two were manual workers. The person who became secretary of the party unit was released from work for two months so that he could attend a course covering the philosophy and goals of the new government. The course also included practical organizational instruction.

In the meeting to officially acknowledge the new unit of FRELIMO, the relationship between these new leaders and the workers was clarified. It was said that the task and duty of the workers is to control the members of the party. Any attitude of arrogance, conceit, any use of party membership as an excuse for getting out of work, or any other form of abuse, had to be immediately reported to the party structures. Disciplinary action would be taken. The party is very uncompromising on this point. The conduct of members has to be exemplary, and it's the responsibility of the *workers* to make sure it is.

Observing this complex process, my political preconceptions about the nature of democratic society were modified substantially.

FRANCES: You must be aware, Peter, that for most people in my country and yours, the very notion of democracy in a one-party state is inconceivable.

PETER: Although I agree that "democracy" is hardly a static concept, I do think that most people in my country and yours, define it. Only then do we have a measure by which to judge our own societies or any others. To me, democracy exists to the degree that there is broad sharing of economic and political power. It assumes, therefore, participation in decisionmaking about how resources are used, as well as a just distribution of the benefits of the development of a society's resources. Do you think we have a democracy in England? Or you in the United States? Does Mozambique have a democracy?

FRANCES: I think you would agree, from your definition, that the verb "have" is not too useful here. The definition that you offered makes democracy not an absolute--today you have it, tomorrow you don't--kind of concept, but a continuum.

In our countries, people confine the definition of democracy to the political sphere. By voting we can participate in selecting political leadership. But our choice is very limited by the fact that those from whom we must choose were selected for us by a system in which wealth plays a large role. Also, by limiting our definition of democracy to the political sphere we resign ourselves to no democratic participation in the workplace and little say in the use and protection of our society's resources.

And, indeed, since we don't include the economic sphere in our definition of democracy, we acquiesce in the tight concentration of economic power--economic power in the hands of a relative few corporations that can play a determining role in the political process. Thus, our failure to include the economic sphere in our definition of democracy means that we risk letting our political democracy become hollow. By this I mean that participation in the form of voting becomes increasingly meaningless because that participation carries with it no power.

PETER: But the task of achieving democracy isn't a simple question of the overnight distribution of power. My experience in Mozambique has convinced me that it is much more complex. Mozambicans seem to be saying that democracy, both in the political and economic arenas, does not spring forth spontaneously and that therefore there must be leadership--some degree of concentration of power--in order to progress toward more democracy.

FRANCES: Concentration of power in order to achieve more democracy? Isn't this a contradiction?

PETER: Well, certainly it can be a contradiction. We have unfortunately witnessed many societies in which concentrated political power acting in the name of protecting the interest of the masses has led to abuse and new forms of domination by a few.

I have come to feel, however, that while any form of leadership can rob people of power, it need not necessarily do so. I have come to see that without leadership, which means some centralization of decisionmaking concerning a society's basic course of development, there is no possibility of confronting the inherited biases in favor of those with wealth, status and education.

FRANCES: Most people are alarmed at the thought of centralization of power. But, let's face it: in every society some people have more power than others. So instead of just asking *whether* power is concentrated, shouldn't we be asking: How concentrated is the power? Are there a mere handful making the society's decisions, as the Portuguese did when Mozambique was a colony? Second, is society moving toward more or toward less participation in decisionmaking? Third, are those with greater power earning it--through their willingness to take on greater responsibility, their competence, and the confidence they inspire in others? And finally, are those with greater power remaining accountable to those whose interests they are supposed to be serving?

These seem like the most important questions.

Peter, what I am learning through your experiences is that it's too simplistic to judge a country like Mozambique to be inherently undemocratic because of its one-party leadership. Instead, we must attempt something much more difficult, something that requires investigation, an open mind and patience--that is, we must evaluate whether the claims of leadership that its power is justified are valid. In part this means determining whether structures exist for popular participation both politically and economically. We know, for example, that in in Mozambique structures have been established for the stated purpose of involving people in decisionmaking. There are Production Councils elected in the workplace, and popular elections play a large role in selecting members of the People's Assemblies, governing bodies on the village to the national level.* But the next step in evaluating democracy in Mozambique is asking whether these struc-

---

* In Mozambique there are nearly 900 local assemblies (with a total of more than 22,000 deputies), over 100 district assemblies (with about 3000 elected depu-

tures really work, whether they really are involving more and more people in decisionmaking.

PETER: It seems to me that the claims of FRELIMO are valid only to the degree that it genuinely promotes accountability to those whose interests it claims to serve--that is, the peasants and workers. And, of course, it is too early to make generalizations about how well the structures for accountability are working in Mozambique. The answers to all the questions you've suggested depend on the integrity of leadership, which is largely dependent on the process by which leaders are selected. But having a good selection process, in which people feel free to say whether or not candidates are people of integrity, is not sufficient either. Unless leaders can be recalled at any time they abuse their power, then there is no possibility of building democracy.

*

ties) and ten provincial assemblies (with about 700) and, on the national level, one body of 200 members which enacts all the principle legislation, ratifies international treaties, approves the budget and so forth. Peasants and workers make up the majority in each of these assemblies. For much fuller descriptions of these structures see *Mozambique and Tanzania: Asking the Big Questions* and *A Luta Continua*, both listed among the resources at the back of this book.

IV. Compromises or Retreats?

FRANCES: Now can we shift back to you, Peter? After your computer work in England, trying to adapt professionally to work at CIFEL must have been difficult, to say the least.

PETER: That's another area where I had to, well, not only modify my views but, in fact, turn them upside down and give them a good shake. Before going to Mozambique, I had worked ten years as a systems analyst using intractive visual display terminals on real-time data base production control systems. Well, the situation that I found at CIFEL was just as different from that as you could possibly get. When I arrived, 22 black Mozambican workers who made up the production council completely controlled the factory. There was really no administration; stock control was virtually nonexistent. And there were all the problems I've already described.

Well, you can imagine the challenge of taking apart all of one's previous training and applying one's skills to such a dramatically different context. Looking back now, I realize just how absurd some of the solutions I proposed in those very early days must have sounded.

Even more different than the culture and the degree of organization (or disorganization) was the fact that here, instead of working to develop control and information systems in which a handful at the top of the factory would have total power, as was my previous experience, now I was to look for ways in which the whole workforce could be drawn into the process of decision-making. Particularly, a goal was for workers to be responsible for the area in which they worked.

My particular skill is in planning and controlling production, but like everyone in those early days, I was drawn more and more into the problems of actually achieving any production to plan!

One interesting little example of the increasing convergence in Mozambique between the political and technical came out in the planning meeting to discuss one of the biggest problems in the foundry--the lack of cooperation amongst the workers. I recall one particular example cited: after finishing their work in one section, workers would sit down and not pass over to assist another section. The party made a rather mechanical analysis of their behavior, claiming it demonstrated a lack of consciousness, a lack of mobilization on the part of the party. And we all nodded our heads wisely.

Then the head of the furnace department got up. Physically he was a really impressive character and also had great warmth and dignity. He said that, while he agreed, he felt that the workers' behavior also was related to the bonus system still continuing from the colonial days. This was the first I'd heard of the bonus system! He said that at the end of every month there was always a fight, at times becoming literally a fistfight, about the question of bonus premiums. He explained that during the colonial period certain sections had been awarded a bonus of as much as 1,000 escudos (U.S. $30). When they were earning a total of 3,000 escudos (about $100 a month) being able to make another 1,000 on top of that was enormously important to the economy of the workers' households.

Only a minority of the workers in the department received any bonus at all. Moreover, there were big differences. The section that used machines to mold sandboxes could earn 1,000 escudos, whereas those molding by hand--more skilled and equally heavy work--could earn only 100 or 200 at the most.

Even worse, the section that was preparing sand was paid at a rate that depended on how much sand they prepared. But, the amount of sand they prepared was dependent upon the amount of sand that was used by the molding machines in their process. And the furnace section was paid by the amount of steel that was poured. That, of course, was also dependent on how many boxes had been prepared. How many boxes prepared was dependent on how much sand had been prepared. So the whole thing is an organic relationship of supply and demand from one section to another. The pace of any one section is dependent on the overall level of production.

Yet workers were getting *competing* bonus rates which means that if you worked in a particular section, it would not be in your interest to assist another section that would perhaps be earning a lower rate than you, or even a higher rate, because assisting them would not be gaining anything for yourself.

Thus the worst possible form of irrational, intersectional competition, characteristic of the incredibly backwards Portuguese concepts of how to manage the factory, was still continuing.

After a short debate, the administrative commission said, "Right. We'll abolish the bonus immediately!"

To my surprise, I adopted a different position. Abolishing the bonus system would be a mistake, I said, given the fact that it was so central to the motivation of the workers at that moment. We couldn't just abolish it. And anyway we certainly couldn't do anything which would cut the worker's take-home pay. Whatever we did, if we did it at a stroke, we risked confusing, even more, the development of the workers' understanding.

I proposed instead that the bonus system be completely restructured to be based on the total amount of finished product per month minus the scrap. So if the workers produced 100 tons of the new product, but ten tons were scrap, then they would be rated at 90 tons of production. This way there would also be an incentive to reduce the amount of scrap. I proposed the bonus should be given to everybody equally, even people who previously didn't earn a bonus. Such a bonus scheme at a later stage could be incorporated into the salary structure without making any kind of abrupt change.

Although in principle opposed to a bonus system, I felt that one had to accommodate, to some extent, the actual conditions. And everyone agreed. But as with so many other things, the stage between agreeing in principle that something should be done and actually implementing it was incredibly long and tortuous. This long delay comes about largely because of the inexperience of the administrators and their lack of self-confidence to make initiatives. Perhaps they fear sticking their necks out. But also there's the relentless pressure of day-to-day problems. Plus the fact that the leadership is incredibly over-extended. The senior administrator is also on the National Commission for Development of Communal Villages, also a director of the Bank of Mozambique, and involved in a whole series of other committees within the ministry and the party. The sheer weight of these multiple pressures and this process of struggle which goes on inside these new leaders create a degree of paralysis when measured against what needs to be done. But even so they are achieving a fantastic amount.

So nothing has happened to date. The bonus system inherited from the Portuguese colonialism continues to run and divide the workers and to create problems.

My previously unquestioned political imperatives were challenged in the harsh reality. In this case, while I'd always held that incentive bonus schemes have no place in the transition to socialism, I found myself arguing against the immediate abolition of the bonus system. It's a small example of many of the compromises which, at a much higher level, President Samora Machel and FRELIMO have had to make as they've picked up the reins of power.

One such compromise announced recently had caused a shudder of apprehension among many: Mozambique is now opening a "window to the West," actively encouraging private multinational investment in agriculture, minerals and industry. This, of course, does not automatically mean, as many Western commentators would have us believe, that FRELIMO has done a complete U-turn.

Speaking with his characteristic frankness to a meeting of the Business International Corporation in Maputo in early 1980, Samora Machel said that the role of the multinational corporation in the third world is to undermine the economic and political independence of newly independent states. But he added: "In our case we know what we want--we repeat--we know what we want and how we want it.... There is a place in this country for private foreign investment, providing its activities are in accord with our political line and our development plans."

This is not the first such venture by Mozambique. Mabor, the abandoned General Tire Company plant, was put into production as a result of a contract between the Mozambique government and General Tire. Eventually ownership will pass completely into the hands of the Mozambique government and the contract includes a comprehensive program of training Mozambicans at all levels.

"Aid" and foreign investment carry risks and costs. Multinational capital is one of the thinnest ices to skate on. Its purpose of course, is to make a profit to serve those outside Mozambique. Whether or not Mozambique will benefit depends on whether the country's leadership can effectively oversee the dealings of the foreign firm in the interests of Mozambique. And that, in turn, depends on how effectively the government can control finance, imports and exports, and can mobilize the workforce in such companies to stand up against abuses. But to really control the direction and impact of foreign firms also requires technical expertise. Given that the few technically trained Mozambicans are already spread thin, it is difficult to see where Mozambicans with the technical expertise to control these new enterprises will come from. Also critical is whether contracts with foreign firms include an eventual transfer to total control by Mozambicans, as in the case of Mabor.

The problem, as ever, is trying to make such compromises, and point them in a direction consistent with one's long-range objectives so that they are not sidetracks into blind allies. Hopefully one is saying, "We can't do it all now, but what we can do is this step. It will take us along the road to the objective that we aim for."

And that, after all, has been the experience of FRELIMO--the whole development of what Samora Machel calls the relationship between theory and practice. He has said many times that theory is not simply something taken from books; theory is not something that falls from heaven. And committing heresy in the eyes of all diehard dogmatists, he declares scientific socialism is not a static doctrine; it is a science that is continually enriched by the daily experiences of mass struggle.

During the liberation war, for example, FRELIMO people found themselves suddenly in control of an area, a village, a farm, and then had to come to grips with how to put into practice a system which didn't result in simply a new form of exploitation.

To live in Mozambique, to work in the factory, was to confront everyday all the prejudices I held.

V. Planning

FRANCES: What I'd like to do now, Peter, is address the issue of planning as part of a development strategy. Planning for the whole society must be impossible really at this stage in Mozambique. The first census is scheduled for this summer (1980). No precise economic or demographic information has been available. Yet planning seems to be the key to whether a society's resources benefit a few or everyone.

PETER: First, let's remind ourselves of what Mozambique is trying to replace. In the colonial days the most obscene wealth and the most appalling poverty existed side by side. With development "planned" by multinational food processors, private raw material extraction industries, and rich landowners and merchants, the spinoff, or trickle-down, is virtually nothing at all to the poor landless worker or urban slum dweller. Now Mozambique's goals of both central planning and worker control are to change all that --to create an economy to benefit the workers. And despite the lack of statistics, and all the other handicaps, very useful steps have been taken to orient production toward the clear needs of the people. Of course, as you point out, a truly planned economy is simply not yet possible.

FRANCES: Let me interrupt you here. I can accept what you say about development in capitalist countries like pre-independence Mozambique,or Kenya or Pakistan today. But some people say that in any socialist economy, *because* it is planned, an army of faceless bureaucrats inevitably develops who serve the interest of a new elite. This is seen as a contradiction to any notion of workers' control.

PETER: Let me say, at once, that the socially planned economy can mean exactly that. It can result in a powerful state elite, an elite that will give itself, for example, housing credits to build houses (which only it can afford), using a capital-intensive building industry that provides little work for people, while the poor and unemployed live in squalor on the other side of town. I'm sure this occurs in third world countries, and often in the name of socialism, or national liberation.

FRANCES: With so many corrupt regimes trying to hide themselves from the people behind progressive-sounding slogans it isn't surprising that there is so much confusion and cynicism about "planned" economics. So what is the difference in Mozambique?

PETER: The first difference I've already spoken of--what products the society places a priority on producing. In Mozambique they are no longer products for the colonial class which has the money, the target group for most investment and production in the third world countries. Instead, production is in line with the plan for national reconstruction put forward by FRELIMO, which gives priority to the needs of the poor. Planning has to be especially rigorous when capital to invest is so scarce, and foreign currency at this moment even scarcer.

It would be ludicrous if, in the name of the free market, foreign currencies so desperately needed to buy spare parts and re-equip the economy were squandered on luxury goods. It would be ridiculous if, because of the shortages now in Mozambique, CIFEL produced nonessential consumables to sell at the greatest profit to the dregs left from colonialist time. Yet, if in fact the capitalist imperative operated in CIFEL, that is precisely what would be happening. In the old days, CIFEL never produced animal-traction plows nor the big African hoe, so much prized by the peasants. Now, because FRELIMO has stopped building new high-rise, luxury housing, or indeed, any concrete-style housing, CIFEL can produce those more needed items. (Now the emphasis is on self-help housing in the shanty-town slums that surround Maputo, and here the state provides much of the more traditional house-building materials. The people supply the labor voluntarily, and the party provides the organizational ability.)

And we shouldn't underestimate the effect on the workers of this change in what CIFEL produces. Workers see themselves as very important people doing tasks which will affect the whole course of development of their country.

FRANCES: Is that your sense of it, or did the workers talk that way?

PETER: Workers genuinely talked in just those terms.

FRANCES: But if in a centrally planned economy the workers carry out plans made at the top, how is it different from just following the directives of the old bosses--except, of course, as you underline, the products are very different?

PETER: While on a national level there can be central planning--meaning guidelines for what should be produced based on what is needed--successful planning depends on the involvement of people at the workplace itself. For ultimately only they know how much they can produce and what they need to be able to produce it. This information is essential to national planners. At its best, then, planning is a dynamic interchange between the workplace and those responsible for national planning.

But to really answer your question, let me tell you more about how planning evolved within the factory.

I participated in two attempts at planning. The first, at the end of 1977 to make the plan for 1978, and the second, at the end of 1978 to produce the plan for 1979-80. The first of these efforts was rather bad. Well, I suppose it had to be bad. None of us had any experience in making a plan. Recall that during this first period, the young opportunists were still in control of the dynamizing group and the administrator was the one who owned two factories and wanted to run CIFEL just as he'd always run his own factories. He considered planning to be nothing more than state interference.

In December 1977, a national commission was formed to try to make a production plan for the country. It was quickly realized that the conditions simply didn't exist for making a plan. Data existed neither about production potential nor the needs of the various sectors of the economy. Even if concretely, say for the construction industry, one could identify the needs of the country--so many bridges, so many hospitals, so many schools--and get from that point to how much raw material would need to be imported, there would still be the question of what it would cost in foreign currency, and how that should be related to the total picture of foreign trade.

Because there was so little data at that time, foreign currency wasn't identified as the limiting constraint, nor the foreign currency divided rationally in any way. The day-to-day consequences of this were enormous. There developed long waiting lists of requests for letters of credit from the state sector controlling importation. Each one was treated *individually*, consuming enormous amounts of time and leading to hopeless crises in procuring imported materials.

Very quickly the plan idea was scrapped and the concept of a program substituted. The difference is this: a program is an indicative process which tries to draw up a general perspective on needs and goals. A plan is a higher state in which specific targets are set.

But back to the factory again. The Ministry of Industry called for a commission at the factory to prepare a plan. The ministry appointed four members of the 12-person commission: me, an East German economist who'd had considerable experience in his own country, a young Portuguese who worked in the factory and the administrator--the first administrator, the one who still had his own two factories. (In the end the young Portuguese was dismissed at the initiative of the workers who felt that his whole manner and approach to work was completely unacceptable. He was arrogant and unsympathetic to what the workers were trying to accomplish within their Production Councils.)

The ministry also called for the Production Council and the dynamizing group to each select four of their own members to participate. Formally, the four appointed members of the commission were a minority; the worker-selected members were the majority. But in practice what happened was that the dynamizing group and the Production Council were assigned the task of analyzing the workforce--how many people were working where--and tasks like that. Gradually they stopped attending meetings. Although they were called many times to participate in the meetings, they never did. Either they had gone to do tasks outside the factory, or they said they wanted to concentrate on doing their specific jobs.

Thus the workers' level of active participation in the planning committee was almost nil. Also, there was little debate or discussion. Planning, such as it was, was very much a technocratic activity.

Nevertheless, considering all the factors, the first plan the committee produced was rather good. It contained a series of production targets and ways in which these production targets were to be achieved--a goal for reducing the scrap and lost time, for example, and what effect that would have on the level of production. For each department, technical measures were recommended to achieve these production levels, including things like the complete overhaul of this or that machine, the installation of machinery that had been bought before independence but never installed, etc.

All these suggestions were theoretically discussed in meetings of the workers in each department. But what tended to happen was that first the proposals of the planning commission were read to the workers. The workers were then asked to comment--to report on problems of production that they'd experienced, particularly over the last year, and to make suggestions as to what should be done. But the workers rarely spoke up.

At the end of this process, it was proposed that we call a mass meeting to approve the plan. The East German opposed this. "What if the workers don't accept our plan?"--a clear indication of his conception of socialist planning. The meeting did take place but, given the makeup of the commission that called it, the meeting was pretty abortive. The dynamizing group secretary displayed the most obnoxious side of his demagogic personality. He read the whole plan and then, in very patronizing language, sought to try to explain it. As he reached the section on the rolling mill he would say things such as, "Who is here from the rolling mill? What do you think about ...?" in a very commanding and insensitive way. And the workers would respond with something such as, "Well, I don't know, I don't work in that section."

"You don't work in that section?" he would ask, and then get very angry. "You're responsible for the whole thing, we're all responsible for everything." He set a tone to the meeting which is very difficult to describe, but in which there could never have been any free exchange of ideas or anything, really. The plan was essentially rubber stamped by this meeting.

After all this, you'll probably be surprised when I tell you that CIFEL's plan was considered to be one of the best, if not *the* best, submitted to the Ministry at the end of 1977! What characterized this plan? Complete inexperience on the part of everyone concerned as to how to set about doing a plan; secondly, a complete lack of organized data of what had happened at the factory; thirdly, a technocratic approach--a mentality of just formally going through the motions of getting the approval of the workers, which ended up making the plan very distant from the workers' experience.

Unfortunately, that's what passes in many countries for socialist planning.

FRANCES: Earlier you described how much attitudes changed in the factory after the dynamizing group was disbanded, the new administrator appointed and the party unit formed. Was the planning process affected, I hope?

PETER: Yes, the new, young Mozambican administrator was very enthusiastic and very committed to the idea that the workers must develop their own self-reliance. So, in September 1978, we were in a very different position to prepare for the planning campaign for 1979.

First, we had a good critical discussion--the administrator, the Production Council, the party and I--about how to identify some of the conditions we wanted to change in order to achieve a different kind of planning process--one which genuinely involved the workers.

The next step was to form a planning nucleus of workers selected by the administration based on recommendations from the party, the Production Council and me. Twenty-four were selected. Four were office workers and the others were manual workers who had shown themselves capable of making a contribution within their field of experience.

At the first meeting of this group we read through point by point the plan that had been made in 1977. We discussed how the plan had been made, and why the various proposed technical measures had not been realized in the vast majority of the cases. If they had been realized it was only very late.

The way in which we organized the reflection process was very good. There was an incredible openness on the part of the administration towards me and the kind of contribution that I could make. We worked in a mutually supportive way. I would write a draft of a discussion paper for the planning nucleus. The first one I wrote was called "Socialist Planning: what's the difference between socialist planning and capitalist planning?" It discussed the differences between the process in which many private enterprises prepare their budgets for the following year and the kind of popular power perspective that people like Marcelino dos Santos, then planning minister, had for the planning process.

At the end of meetings discussing such topics, I would write up the points discussed and the measures proposed. These decisions made up an embryonic planning document. The emphasis throughout the planning process was the question of liberating the creative potential of the workers by creating conditions in which *they* could make the maximum contribution.

Some surprising insights came out of these initial discussions. We got into a discussion about the amount of scrap--wasted raw material generated by the rolling mill. Theoretically, in a mill of this kind, scrap should be around three percent, but by 1976, the percentage had soared to about 30. This meant that the cost of the steel bars we produced was considerably higher than the cost would have been if, instead of raw materials, we had bought the finished product from South Africa!

We discussed how to reduce the scrap. We analyzed the various points in the production that generated scrap. Our analysis showed that a particular bar-cutting device had broken a couple of years ago and that, as a consequence, whenever a roller jammed towards the end of the rolling process (you can visualize the mill as a continuous process of rollers ever decreasing in diameter), everything that was following behind had to be cut out of the machine and scrapped--an enormously time consuming process.

When we analyzed how that could be avoided, someone said, "Well, of course, in the old days we used to deal with the problem by deviating the bars in process to another part of the machine where they could be turned out into a finished product of a different size. If we were making 6 mm bars and the finishing roller at the end of the line jammed, instead of having to wreck all the bars that followed, this deviator was switched on to allow the mill to produce 12 mm straight bars. All the steel that followed was saved."

I couldn't believe it! Why, I asked myself, did this critical piece of information emerge only at this point? Why, when the Production Council was meeting every Friday to discuss the problems in the mill? Why, when the electrical department had weekly meetings to discuss the problems of maintenance? Why, when the Portuguese technicians were involved and were capable of writing an internal memo giving instruction for this repair to be made?

The answers to these questions are very complicated. Achieving a factory that operates as an organism which regulates itself implies certain communications and control systems, certain levels of authority and responsibility. With the exodus of the Portuguese, an alternative structure was created, but *not* a communication system which acted in an organic way. This is a problem of workers' control: how to get beyond just being a talking shop to becoming a self-regulating mechanism. It is this aspect, the organizational theory about what a worker-controlled firm really is, that most fascinates me.

One problem was that the workers were meeting together each Friday without organized data--without data which said where and how much scrap was being generated, how much machine time was being wasted, what the daily production had been, and why there were variations in daily production. The workers would come to the conclusion that it was necessary to do this or that particular thing, but somehow their decision never got communicated to those who could act.

The Production Council clearly had a role to play in controlling the work flow and the organization in each section. But it would have been anarchy for the Production Council to tell the electricians to stop doing that job and do this or to authorize capital expenditures in isolation. Such measures should have been coordinated by the Production Council at its factory-level meetings and then passed sideways to the administration. The point, of course, was that under the old administration, the old administrator would neither encourage nor welcome such participation. The translation of points discussed to changes within the factory never occurred.

FRANCES: You said that in September 1978, the planning nucleus started trying to create the conditions in which a realistic plan could be formulated for 1979--a plan in which the workers' creative capacity could be expressed. What specifically did the workers' planning nucleus change about the process of planning?

PETER: First of all, planning had been seen as a *one time* activity. The plan had been produced but with no mechanism to control its implementation. There was no effort to communicate even the overall targets in a comprehensible form to the substantially illiterate workforce. Certainly that had to change.

Also we figured out why workers hadn't discussed certain factory problems in the past. Asking very general questions, such as "What are the problems in your sections?" hadn't stimulated workers to really examine their situation. We decided more specific questions were needed.

To stimulate worker discussion, we prepared a checklist of very specific questions such as: Which machines have broken down over the past year? Which are in need of a complete overhaul? How is the raw material situation? What stock control routines exist? A whole checklist like this was read aloud at department meetings two weeks before the meetings were planned for the sections.*

*There are three departments. In each department there are 6 to 20 sections such as furnace workers or carpenters.

The representative of the Production Council stressed that this checklist wasn't for discussion now. Nevertheless, it was interesting that during every meeting in every department, when that list was read, the workers started to discuss the questions spontaneously. "Yes, I wanted to say something about the overhead crane because I think it needs this or that." Finally we had to stop their comments: "No, stop, we will discuss this later. This is just for you to think about now, to discuss within your daily experience, at lunch breaks, etc."

To give an orientation, two or three members of the planning nucleus and a party member went to these first-level section meetings.

As simple as it sounds, our second organizational breakthrough was having one of the more literate members of the planning nucleus take minutes. This was very successful. They took very precise minutes. The earlier experience of writing minutes for the meetings of the planning nucleus had obviously been of use, showing the need and the way to take minutes. The need to learn such a simple procedure, taken for granted among most workers' movements in the world, underlines how Portuguese colonialism stunted Mozambicans.

Monitors from the planning nucleus coordinated these first-level section meetings. Then, at department-level meetings, all of the contributions came together and the general problems of the factory were discussed.

FRANCES: Listening to your story, Peter, one aspect impresses me so much and I am sure it would make many envious: for all the problems and hardships you had to face, you seem to have felt that you were *useful*. In my country so many are denied the right to feel useful. So many are either unemployed or in meaningless jobs.

PETER: I understand what you are saying. Because I did feel useful, the later part of my stay in Mozambique was one of the most satisfying periods of my life. But in the beginning, I didn't know it would go that way. During the period under the first administrator I contributed very little. It was very trying. I felt that as a matter of conscience I couldn't go along with what the administrator was trying to do. But how could I, as an individual foreigner, oppose him when apparently no one else did?

Fortunately, I was in Mozambique long enough to see my work relationships become completely transformed. The new administration was always open. A suggestion would often lead to a discussion until very late in the evening. If I had anything which I thought I could contribute, the administrator would always respond, "Write us a short paper on that and we'll discuss it as soon as possible."

FRANCES: What activity gave you most personal satisfaction?

PETER: The formation of the planning department. Recall that one of the conclusions of the second planning campaign (the one at the end of 1978 which planned for 1979 and 1980) was that we had previously failed to make planning a day-to-day activity. Planning was a *document* locked away in the drawer instead of a dynamic activity that responded to changing conditions. Planning reflected the kinds of relationships within the factory at that time: it was not only a document but a dead document, a symbol of the bureaucratic way of working that characterized the old administration.

Another problem which we identified was that, although the Production Council was meeting every Friday to supposedly discuss the problems of the week, it was not armed with any statistics. No one even knew how much had been produced that week. No one knew how much time had been lost, how much scrap had been produced, what the major breakdowns had been, how much raw material had been consumed, and so forth.

But in the second planning campaign we identified these problems and, during the discussion, it was agreed that we would form a planning department. We drew together three manual workers, two clerical workers and two young students. For three months I organized a course every afternoon on principles of planning. I can never remember enjoying my work in any place that I've ever worked as much as I did those three months at CIFEL.

The planning department started off by discussing methods to graphically represent data. The first example which I presented was of six hens laying eggs, but not every hen every day. One big square of graph paper represented one egg and the workers had to plot 20 days' production from these six hens. That graph took 3½ days to produce! Even the drawing of the lines for each column was a major task for some of the young workers, but we did it. By the end of the course we developed a very satisfactory way of presenting the production data for the factory.

We divided the total production for the year into its 12 monthly blocks. We said that since there would be improvements made over the year, production later on in the year should be higher than at the beginning.

So we colored each month with a separate color. Another column of 12 colored blocks alongside represented the actual production that we achieved. We could see where we were relative to the year's total target and we could see the performance of each month compared with the month's target. All this was very successful. It was very easy to show to meetings of all the workers. It served as a kind of thermometer throughout the year, relating actual production to the targets. The production targets were no longer a dead plan but a living thing which the workers could understand and control.

The next thing we tackled were weekly graphs for each machine or section, showing what had actually been produced each day over the last six work days. Underneath was another graph which showed how much scrap or wasted work had been produced as a percentage of the total. Another graph showed the number of hours worked per day, with a red block showing the amount of time lost due to mechanical or electrical faults, etc.

So you can see we achieved an information reporting system specific to each section and even each machine. Each worker or work group had a folder in which each week's graphs were put. On the daily production graphs the workers drew a line representing the best production achieved daily during that year. That level then became the new production target which they would try to beat! Thus the rather vague idea of emulation (the socialist alternative to individualist competition) was given concrete form in objectives which workers could relate to. Competition is important: it's the motor of capitalist economy. This new form of competition, competition to contribute more, is a very important idea but incredibly difficult to implement.

We also worked on the theoretical capacity of the machines. The workers found this concept incredibly difficult to grasp--for example, the idea that if one machine produced five tons and those five tons passed to another machine for finishing, then in fact the finished product at the end of the day would be only five tons and not ten tons. Part of the problem was the difficulty of dealing with abstract concepts. The Portuguese education system did not deal in abstract concepts at all; mathematics was very mechanistic and rote-learned, completely divorced from the real world. Therefore, representing abstractions like "capacity" was incredibly difficult. Learning to calculate percentages took weeks, literally weeks, of effort on all our parts.

But what was different and important about our work group was that we were self-regulating. We refused to allow a hierarchy to be established. It was made very clear that I would soon be leaving and that the group would soon be responsible for carrying out all of these tasks. To insure that the group remained in touch with the grassroots reality, we all worked one morning per week doing manual labor on the shop floor.

A lot of the training centered around initiative taking, individual leadership, individual responsibility. This was the area that consumed an enormous amount of emotional energy of all of us. I would propose tasks for the next day and nominate someone to be responsible for coordinating the work. Very often, when I would arrive in the morning, the work would not have been started--not because they *couldn't* but because they couldn't convince themselves that they could. They had been taught that they couldn't. We spent more time in group analysis of why this was, what each of us had done, what each of us should have done. They were really criticism/self-criticism sessions, which often went on after working hours. Gradually workers came to be able to do all of these things.

There's another very unpleasant but very revealing story I'd like to tell you. As part of the investigation of the theoretical capacity of the foundry, I arranged for members of our planning group to meet with the Portuguese technician in the foundry. They were to spend the morning with him. When I arrived at lunch time to see how they got on, there was a cloud of depression in the office, eyes down to the floor, long faces. I said, "Okay, how did it go?" There was silence. After a few minutes, one looked up with such indignation in his face and so much tension in his body that the pencil in his hand snapped as he spoke. He went on to describe what had happened. The Portuguese technician had given nothing but a tirade against the workers, saying that they were lazy, that they had the understanding of children, that the Mozambican people were incapable of doing anything, that the economy was falling into ruin, and that if only they'd listen to him, he could solve all of the problems. The effect of this was that all of the confidence building we'd worked on for weeks was shattered in one day!

Although it was time to break for lunch, I went with them to see Pedro, the head of the furnace department who was important to our evaluation of the factory's capacity. We asked him a series of questions about the capacity of the furnace, about the temperature for pouring, things like that--all questions which he answered well. We asked his opinion about what needed reorganizing in the department, and information about procedures used during the colonial period which were not being used now, all of which he answered fully to the best of his experience. Gradually, as we talked, I saw a smile break on each of the young Mozambicans' faces. We went back to the office and analyzed why the Portuguese technician had said what he'd said, why it upset them so much, and what the objective reality was.

That afternoon we measured the foundry floor to calculate how many molding boxes it would be possible to cast in a day. We estimated how many molding boxes could be produced by the molding machines and how much sand that would require from the sand preparation department. From that we identified where the bottlenecks would be. The group presented a short report on all this to the administrator, who was very enthusiastic and supportive.

The workers didn't understand the full details of everything. Certain assumptions had been made which simplified the reality, that's true. But the workers understood that as well. The important thing is that they learned that the process of production can be controlled, can be viewed in a scientific way rather than in the day-to-day, haphazard way adopted out of desperation. They saw the need to plan, the need to control and, most importantly they felt capable of doing it themselves.

Just before I left Mozambique, I went for two weeks' holiday in Ethiopia. Returning to the planning department, I was met with the usual warm greeting but I could tell that there was something mischievous in the air. As I was talking, the workers stepped back to reveal on one of the wallboards a dozen graphs which they'd produced during the two weeks I'd been away. The graphs consolidated everything we learned together in the course. After I left, I received letters from them and from people who had visited the factory. They said that the planning department continues on. It was very active in producing the plan at the end of 1979.

The work with the planning department really is, I think, the most useful thing that I contributed in the two and a half years I was there.

VI. Reflections

FRANCES: But Peter, since many people view human nature as inherently selfish, they believe people would have to be forced, or at least coerced, to cooperate and to share. So often I hear the attitude expressed, "What the hell, workers don't give a damn about anything but their paychecks anyway."

So, naturally people are suspicious of any movement that claims to be based on the principle of cooperation, rather than just self-interest. They feel that Marxist regimes, in forcing people to behave in a manner alien to their nature, end up as cold and uncaring as the previous capitalist ones. They see people under such regimes as gray and expressionless, with human kindness and love subjugated to the will of the minority--and all in the name of righteous-sounding principles.

PETER: My experience shows that it can be just as bad as many people imagine. It can be, but it needn't be; that is what I've learned. Perhaps this is central to the message I want to express.

Looking around the world we see so much abuse and so much cynical authoritarianism carried out in the name of socialism that it's hardly surprising that the western propaganda machine can persuade us easily that socialism is synonymous with coldness and alienation. Since our media is controlled by those hostile to socialism, we can lose sight of the real gains the working people have made in some socialist countries.

FRANCES: I think that while many of us see the hostile actions of the big powers and multinational corporations against struggling socialist countries, it is hard for anyone living outside such societies to appreciate the enemy *within*.

PETER: Yes, people don't understand the influence of the opportunists, like the young head of the dynamizing group who wanted the workers to look to him, or the influence of those left over from the old elite, such as the first administrator. Sure, if people such as these come to power, what *can* one expect? New relationships couldn't flourish in the kind of environment they created, only a duplication of the relationships of the period they came from. But is this the failure of socialism? Or the failure to achieve socialism?

As I said before, attempts at new social relations can be like you've described, but they needn't be like that. A thousand anecdotes come to mind that illustrate this for me--how factory life changed during the period of the second administration and after the FRELIMO group formed. Let me just share one.

The new party group and the new administration decided to commemorate the Mozambican Heroes' Day which at that time was the 10th anniversary of the death of Eduardo Mondlane, the first president of FRELIMO. They organized a fiesta on a Saturday afternoon after work. A committee was formed, and it decided to hold the festival in the heart of the factory, in the main gallery of the foundry. Anyone who has seen a foundry can imagine what a dark, grimy and gloomy setting that must be! The walls are thick with smoke from the furnaces and everywhere there's a fine black dust that comes from the molding sand. So the first task was to paint the walls white. The workers also painted symbols of the new Mozambique, such as peasants planting and harvesting crops. They painted slogans about the continuing sacrifices of the war and the need for each person to contribute to the national reconstruction program.

The opening speech was made by a young person from the Ministry of Industry. I recognized her as being the typist who used to prepare my paycheck at the Ministry, a person one wouldn't notice again perhaps, but here she was now at the factory, surrounded by 400 working men, her small frame insignificant alongside the great, looming machines. Now she spoke with a confidence and force which commanded respect and attention. It's incredible to think that that mother was almost certainly bought as a wife, perhaps a second or third wife, to work on the husband's plot of land. And here she was, a leader of men, a dynamizer of the new revolution, a symbol of the new Mozambique.

Next followed the first performance of the new popular culture groups. We'd all been looking forward to this day. We'd seen them practicing around the factory. But this was the first time they would come together to give a performance.

Popular culture in Mozambique is a difficult problem. Many of the original dances of the pre-colonial days have origins in initiation rites, tribal conflict and the like--patterns inconsistent with the new Mozambique. So these dances, which maintained more or less the traditional forms, were now given new content.

The first dance was a spectacular acrobatic dance that involved a lot of "limbo" style body stretching. The next two dances were from the south of Mozambique and involved lining up and performing a series of ritual steps. Out of the line emerges one figure who, with a series of stylized dance routines, proceeds with a chant much appreciated by the audience. The first dance satirized the administrator. It was obviously hilariously funny, but no one would translate it from the tribal language for me!

The second dance was led by Pedro, the leading furnace man. He has enormous height and striking physical appearance but incredibly gentle eyes and sensitive manner. His contribution was in Portuguese (so I could understand)--a story of the "flame of revolution" he called it. The flame is in the hearts of the Mozambican people, he said. And it is spreading across Zimbabwe. At that time, Nicaraguan liberation was also very much discussed in Mozambique. He ended up in a very moving and dramatic way with a very passionate statement about the total liberation of all people from oppression, living together in peace. I don't want this to sound as though it was some kind of rhetorical political speech passed off as art. It wasn't that at all. The accompaniment of the gradually increasing rhythm of the dances and the passion and compassion expressed in his words was true art.

The final dance was another acrobatic, a kind of warrior dance which involved kicking up a lot of molding sand and dust, which no one seemed to bother about.

Finally the administrator rose to his feet to make a short address. He started with a very solemn face and turned to everyone and said, "Well, just before we come to the meal, I should remind you that the position of FRELIMO is to oppose drunkenness and alcoholism. That is correct and we support that." There were anxious looks passed from face to face and a kind of rumble went round the room. Suddenly his face cracked into a line of bright teeth. He said, "No one is going to be rigid about implementing that, of course. We're not here to impose dogmatic positions, we're here to grow to understand what these things mean."

Then he went on to make a very beautiful and passionate speech. Looking around at the machinery now lying silent, he said, "During the colonial period these machines had dehumanized man. And the avoidance of work had been the only form of protest that the Mozambican people were able to make. Work

PROLETARIOS DE
TODO MUNDO:
UNI-VOS
LK

was alienation. Now," he said, "we have to defeat that, to understand that work now is for the liberation of people--the liberation of suffering from poverty, from crippling ill health, from old forms of cultural deprivation." Then he turned to the audience and said quietly, "Our struggle is not just a struggle for production. It's a struggle also for making ourselves more human, for creating among ourselves human relationships of a new kind. He said, "Okay, today we might drink perhaps a drop too much, but no one's going to worry about that. What is important is that we use this moment to talk to each other, to say things which perhaps, during the tensions of day-to-day problems, we couldn't say. Perhaps we can use this time to get to know each other, to feel ourselves to be more human and caring."

It was a beautiful speech, for me a very moving experience to be there in that room decorated as it was, with those ordinary working people sitting there, themselves moved by the words, by the warmth of the sentiments. As he finished, he reminded us that this was the first fiesta that had ever been organized in the factory, that it was possible only because of the blood of the Mozambicans that had given their lives in the sacrifice of the armed struggle. Then he said, "Let's celebrate with joy the new possibilities to celebrate with joy, because this is now our factory, and we're able to make it a factory fit for humans to work together in."

He stopped and sat down and there was a great roar from the workers gathered there, not just applause, but a *roar*. Some workers got to their feet and there was a spontaneous shout of "Viva FRELIMO." Not clenched fists going up in a mechanical way in compliance with Party directives. Not brainwashing, not propaganda, not anything which fits our stereotypes, but a true cry from the hearts of those people recognizing something of what they had gained and what the future held for them.

This is a small example. It doesn't say that Mozambique has transformed social relationships, nor that the process is irreversible, nor that progress is uniform across Mozambique.

FRANCES: What is significant to me about the spirit of this festival, Peter, small as it was, is that it points to what could be. Also, it forces me to reflect on the kind of values and sentiments considered normal in an office party in my country. None of these kinds of sentiments would be expressed there.

PETER: I also want to return to the notion of people being gray and inhuman in socialist states. About 120 kilometers north of Maputo there's a place called Praia do Bilene. This is a natural lagoon, clean water, free from sharks, a beautiful clear and clean beach, absolutely safe for swimming. In the old days of Portuguese colonialism this was a famous seaside pleasure spot. The lagoon hummed to the constant roar of ski boats and sports fishing boats.

Now all that's finished. Generally speaking, the people who had the ski boats left. It's very unusual to hear any motorboats at Praia do Bilene. Now there are new sounds. Praia do Bilene is the first *workers'* holiday center. Workers and children from all over the country take holidays there. It's financed and maintained by the state. For a very small fee you can rent a cabin there. The glamor and elite image of Praia do Bilene is gone. The sound of the motorboats is replaced by the laughing of children who for the first time in their lives are swimming in the sea, playing football on the beach, and involved in gymnastic and cultural dance activities organized by FRELIMO people at the resort hotel.

FRANCES: I suppose that whether one sees the new Praia do Bilene as "gray" because it is less glamorous or whether one sees the change as something beautiful depends very much on the eye of the beholder.

PETER: Is a society that doesn't have racing cars and "Hollywoods" gray? Perhaps we should ask gray to whom.

As for the suggestion that workers are really not interested in anything but paychecks, I've already mentioned the group of workers who helped to harvest rice. This is voluntary labor for which they got absolutely *nothing*. How does such a view of people account for workers like these? They got up before dawn, walked many kilometers, caught the bus for a two or three hour drive, worked a full eight-hour shift, then voluntarily attended literacy classes. Some of these workers were 50 or 60 years old.

FRANCES: Your experiences in Mozambique seem to contradict all those who want to present the human condition as fixed and simply individualistic.

PETER: Many people *are* cynical and disaffected under capitalism and under much of what is called socialism. It's easy to understand disaffection in the factory during the period of the first administrator. But to generalize and say that is the human condition is to belittle the dedication of many Mozambicans and to reinforce cynicism where there could be hope.

All the workers at CIFEL voluntarily contributed one day's pay per month to the Solidarity Bank, used to support liberation movements and victims of national disasters (like the war and floods). I attended meetings to remind workers of the importance of this fund and to reaffirm support for it. I felt that workers genuinely supported the Bank and that if there had been significant opposition to it, the Bank would have been killed. I really enjoyed some of those meetings. To hear ordinary working people with very little money for their own families talking in such an animated way about the needs in Zimbabwe and Nicaragua is pretty inspiring stuff.

FRANCES: What were your feelings when you returned to England? Did your views change after achieving some distance from your experience?

PETER: Being out of England for two and a half years was too long without a break. It meant that when I returned to England I underwent an acute cultural shock. I stayed with some people who had a young son and daughter who had left school, were unemployed, with no prospect of finding work, and were very disaffected and cynical. It saddened me to see young people who had no view of what they wanted to do or any optimism for the future--either their own future or the future of the country of their birth. How they contrasted with the young Mozambicans that I'd just left! I found that difficult to cope with.

Then there was the overwhelming reality of materialism in our society to adjust to again--things which for two and a half years I had managed without, and not missed, were once again being thrust upon me as essentials. To walk around a typical English supermarket after what I'd seen in Mozambique was obscene and depressing but at the same time tantalizing.

Other small things hit me: my children, instead of singing the FRELIMO songs of optimism, construction and working together, were now singing the latest television commercial jingles.

Maybe that's a small thing, but together with that is the incipient sexism they brought back from school and the barrage of media portrayals of the Patriotic Front in Zimbabwe as bloodthirsty terrorists. This was completely contradictory to everything that my children had actually experienced and come to feel and to believe. Worst of all, I think, was the trivial nature of life.

To conclude, I want to make sure that my experience is seen in perspective. Please keep in mind that CIFEL probably is an exceptional enterprise in Mozambique--exceptionally good in terms of the political values we have focused on--even with all its obvious problems. The second administrator was certainly a person of exceptional sensitivity and political awareness. But just how exceptional my experience was is hard for me to assess. I *do* know that unless the entire society were supporting the kind of efforts we were making within CIFEL, the progress we did experience would not have been possible. While exceptional, I am sure that my experience was not unique.

The point that I've tried to make, of course, is that Mozambique is *not* some kind of workers' paradise. How could it be with all the problems that it faces amidst conditions which are impossibly bad for even maintaining itself, let alone creating the new society Mozambicans envision?

I think mistakes have been made; incorrect positions have been taken. For example, abortion is still illegal. I don't like certain trends with regard to salary differentials, a matter which is still unresolved. I don't like a certain tendency towards exalting bureaucrats and standing when a department head enters, and a recent statement that the wives of ministers should not be in the same hospital ward as the wives of cooks. But for each of these forboding examples one can find another position stated. For example, as a counter to the tendency of giving privilege to bureaucrats, there is intense discussion about the role and power of the popular committees. It is a process of discussion, a process of struggle--of struggle between those who want to develop a truly popular, democratic society, in which ordinary people participate and control.

While such struggle continues, victory by either side is uncertain. More mistakes will be made, more retreats will take place. What stands out in my mind is that Mozambique is not a workers' paradise, not a model for all third world societies, or anything crass and simplistic like that. But its experience shows some very positive trends. It's this that I want to express again and underline. Words like "socialism," "planned economy," the "Party" and "the revolution" are not single-meaning stereotypes; they are the product of struggle, of conflict between the old and the new. I've seen enough to be able to say with complete honesty and confidence that for all of the bad models one can find, one can see alternatives, glimpses of how it could be.

Conclusion

CONCLUSION

We believe the workers at the CIFEL steel mill, along with many of the 12 million other people of Mozambique, are striving to create a nation where everyone has a voice in making the policies--and sharing the products--of the whole society. Sadly, there are few countries in the world today where so many are engaged in such a challenge. These few deserve our special attention and our special support.

We who live outside these countries must ask ourselves: What can we do to make our own governments' policies less of an obstacle in the Mozambican people's struggle for true independence and a democratic society?

In 1980 Mozambique will be able to grow only one-third of the food it needs. A drought wiped out the entire corn crop in southern Mozambique in 1979. Rhodesian raids aimed at crippling Mozambique's support for the Zimbabwean freedom fighters disrupted food production in the prime agricultural area, the Limpopo Valley. The raids also damaged the rice fields and will reduce the 1980 rice harvest.

Mozambique needs help and is asking for help. But because we know the dangerous impact of foreign assistance from most Western governments, we must lobby for specific kinds of aid to Mozambique. Debt cancellation grants of foreign exchange or food with no strings attached are what Mozambique most needs. Otherwise, aid programs from governments unsympathetic or hostile to Mozambique's social and political goals could well subvert its direction. Americans should also keep abreast of attempts in Congress to curtail or restrict American aid to Mozambique.

As individuals, we can directly help to build the new Mozambique by sending contributions to the Mozambique, Angola, Guinea-Bissau Information Center (MAGIC), earmarked for its "solidarity fund." This fund is used for development programs in Mozambique.

We can also contribute to the many support groups, such as MAGIC or the Toronto Committee for the Liberation of Southern Africa, that educate North Americans and Europeans about the struggle and needs of the Mozambican people. At the end of the book we have listed those Africa support groups with which we are in touch. But there are many others. So check also in your own city.

Mozambique also needs skilled *cooperantes* willing to commit themselves for two years of work. Medical personnel, civil engineers, and teachers (especially in the sciences) are especially needed. Interested Canadians should write to John Saxby at CUSO, 4824 Cote des Neiges, Montreal, Quebec H3V 1G4. Presently there is no structure in the United States for recruiting and screening *cooperantes*. U.S. citizens interested in becoming *cooperantes* should write to MAGIC only if they would be willing to go to London for interviews.

Equally important, we can ally ourselves with the efforts of the Mozambican people by working to expose--and end--the support which our governments and private corporations based in our countries give to the apartheid government in South Africa. Neither Mozambique nor the new Zimbabwe will have full freedom to develop as long as they face naked hostility, and a growing guerrilla war, from neighboring South Africa, where four million whites rule over 22 million blacks. By pressuring our government and our banks and corporations to end their support for South Africa, we can bring the day of majority rule there closer, and so aid the Mozambican people, too.

These are some of the concrete steps that we can take to give the positive initiatives in Mozambique the best chance of growing. But the challenge presented by the CIFEL workers can also reach into our daily lives. For, in the last analysis, we will not believe that the changes sought by the workers at CIFEL are possible unless we can experience ourselves changing. Observing the courage of Mozambicans to sacrifice and to struggle for change should challenge us to do likewise.

--To confront our fixed, and perhaps too narrowly defined views, of what makes a society really democratic.

--To question whether our daily life choices--where we work, where we shop, where we bank, what we teach our children--support a system hostile to what the Mozambicans are attempting.

--To join with organizations locally and nationally confronting the concentration of power and wealth in our societies and working to democratize our economic and political life.

--To develop the patience needed to maintain our hope over the long period necessary for genuine change, change not only in structures, but in the attitudes and values of people.

And finally, the Mozambicans challenge us to counter despair and cynicism--perhaps the greatest obstacles to change. We cannot do this by painting idealized visions of political and economic alternatives, but by discovering how to spread the word about courageous efforts of "ordinary people" throughout the world seeking dignity and self-determination.

RESOURCES

PERIODICALS AND BULLETINS

1. *Africa News*, P.O. Box 3851, Durham, NC 27702 (weekly, $20 per year, individual).

2. *AIM* (monthly information bulletin; available in English), Mozambique Information Agency, Avenida Ho Chi Min 103, Maputo, Mozambique. ($10 per year).

3. *Review of African Political Economy*, Merlin Press, Sufferance Wharf, 2-4 West Ferry Road, London E14, England. (Three issues per year, $5 per year).

4. *Southern Africa*, Southern Africa Committee, 156 Fifth Avenue, New York, NY 10010. (Bi-monthly, $8 per year).

5. *Tiempo* (Mozambique weekly periodical; in Portuguese only), C.P. 2917 Maputo, Mozambique.

BOOKS

1. Allen Isaacman, *A Luta Continua*. Fernando Braudel Center for the Study of Economics, Historical Systems, and Civilizations (State University of New York at Binghamton; Binghamton, NY 13901) 1978. Excellent summary of Mozambique since liberation.

2. __________, *The Tradition of Resistance in Mozambique* (University of California and Heineman: Berkeley and London) 1976.

3. George Houser and Herb Shore, *Mozambique: Dream the Size of Freedom* (The Africa Fund, 1964 Madison Avenue: New York, NY 10016) 1975.

4. Frances Moore Lappé and Adele Beccar-Varela, *Mozambique and Tanzania: Asking the Big Questions* (Institute for Food and Development Policy, 2588 Mission St., San Francisco, CA 94110) 1980.

5. Samora Machel, *Mozambique: Sowing the Seeds of Revolution* (MAGIC 34 Percy Street, London W1P 9F6, England) 1974.

6. Eduardo Mondlane, *The Struggles of Mozambique* (Penguin: Baltimore and Middlesex) 1969.

7. John S. Saul, ed., *The State and Revolution in Eastern Africa* (Monthly Review: New York and London) 1979.

ORGANIZATIONS

1. Africa Resource Center, 464 19th St., Oakland, CA 92612. (415) 763-8011. Library open to the public. West Coast distributor of publications of the American Committee on Africa. Speakers available.

2. American Committee on Africa, 198 Broadway, New York, NY 10038. One of the oldest U.S. organizations supporting African liberation.

3. Liberation Support Movement (LSM) Information Center, P.O. Box 2077, Oakland, CA 94604. West Coast resource on liberation movements, particularly in Southern Africa.

4. Mozambique, Angola, and Guinea Information Center (MAGIC), 34 Percy St., London W1P 9F6, England. Publishes "People Power," a bi-monthly information pamphlet on the three liberated countries as well as documents in translation. Also coordinates program to screen and send skilled "cooperantes" (helpers) as needed by the three countries.

5. Southern Africa Media Center, California Newsreel, 630 Natoma Street, San Francisco, CA 94103. Comprehensive library of films on Southern Africa available in North America; offers rental policy for needy groups.

6. Toronto Committee for the Liberation of Southern Africa, 121 Avenue Road, Toronto, Ontario, Canada M5R 2G3. Publishes a newsletter and occasional papers.

7. Washington Office on Africa, 110 Maryland Avenue, N.E., Washington D.C. 20002. (202) 546-7961. Newsletter alerts readers to U.S. policies affecting Africa.

ABOUT THE INSTITUTE...

The Institute for Food and Development Policy is a not-for-profit research, documentation and education center. It focuses on food and agriculture, always asking: Why hunger in a world of plenty?

By working to identify the root causes of hunger and food problems here and abroad, the Institute provides counter messages:

* No country in the world is a hopeless basket case.
* The illusion of scarcity is a product of the unequal control over food-producing resources; inequality in control over these resources results in their underuse and misuse.
* The hungry are not our enemies. Rather, we and they are victims of the same economic forces which are undercutting their food security as well as ours.

The staff of the Institute: Frances Moore Lappé and Joseph Collins, co-founders, David Kinley, Bruce Randall, Douglas Basinger, Gretta Goldenman, Jennifer Lovejoy, Chris Anderegg, Stephen King, Regina Fitzgerald and Wendy Ward.

FINANCIAL SUPPORT

The Institute solicits contributions from individuals, church groups, and private foundations. In addition, monies are received through speaking engagements and literature sales. The Institute does not accept funding from governments or corporations.

More and more, the Institute depends on individual contributions, literature sales and speaking engagements. In addition, we are grateful for current donors: Columbia Foundation, W.H. and Carol Bernstein Ferry, Oxfam-America, Gretl Popper, Rubin Foundation, Seed Fund, Tides Foundation, The Youth Project, Agricultural Missions, Church of the Brethren, Maryknoll Fathers, Presbyterian Church in the U.S., United Methodist Church, United Presbyterian Church of the U.S.A., and First United Church (Oak Park, Illinois). Financial support does not necessarily reflect agreement with the views expressed.

FRIENDS OF THE INSTITUTE

Because the Institute's work threatens many established interests, we believe that our effectiveness depends on developing the widest possible base of support. By joining the Friends of the Institute program you can receive our expanding list of publications at a generous discount or free.

All contributors of $25 or more receive a free copy of the paperback edition of the highly acclaimed *Food First: Beyond the Myth of Scarcity* by Frances Moore Lappé and Joseph Collins with Cary Fowler (Ballantine, 1979). Contributors of $100 or more also receive one free copy of all major Institute publications for one year. Contributors of $25 or more also receive a 50 percent discount on one copy of all Institute publications for one year.

All contributions are tax-deductible.

INSTITUTE PUBLICATIONS

World Hunger: Ten Myths clears the way for each of us to work in appropriate ways to end needless hunger. Frances Moore Lappé and Joseph Collins, revised and updated, 72 pages with photographs. $2.75

El Hambre en el Mundo: Diez Mitos, a Spanish-language version of *World Hunger: Ten Myths* plus additional information about food and agriculture policies in Mexico, 72 pages. $1.45

Food First: Beyond the Myth of Scarcity, 50 questions and responses about the causes and proposed remedies for world hunger. Frances Moore Lappé and Joseph Collins, with Cary Fowler, 620 pages, Ballantine Books, revised 1979. $2.95

Food First Resource Guide, documentation on the roots of world hunger and rural poverty. Institute staff, 80 pages with photographs. $2.95

Aid as Obstacle: Twenty Questions about our Foreign Aid and the Hungry demonstrates that foreign aid may be hurting the very people we want to help and explains why foreign aid programs fail. Frances Moore Lappé, Joseph Collins, David Kinley, 200 pages with photographs. $4.95

Needless Hunger: Voices from a Bangladesh Village exposes the often brutal political and economic roots of needless hunger. Betsy Hartmann and James Boyce, 72 pages with photographs. $2.95

What Can We Do? An action guide on food, land and hunger issues. Interviews with over one dozen North Americans involved in many aspects of these issues. William Valentine and Frances Moore Lappé, 60 pages with photographs. $2.45

Mozambique and Tanzania: Asking the Big Questions looks at the questions which face people working to build economic and political systems based on equity, participation and cooperation. Frances Moore Lappé and Adele Beccar-Varela, 128 pages with photographs. $4.75

Agrarian Reform and Counter-Reform in Chile, a firsthand look at some of the current economic policies in Chile and their effect on the rural majority. Joseph Collins, 24 pages with photographs. $1.45

Aid to Bangladesh: For Better of Worse?, an interview showing the misuse of aid throughout Bangladesh. First of Impact Series with Oxfam-America. Michael Scott, 28 pages with photographs. $1.45

Reprint Packet No. 1, a compilation of eight articles by Institute staff which have appeared in a wide range of periodicals, 16 pages. $1.45

Research Reports. "Infant Formula Promotion and Use in the Philippines," Frances Moore Lappé and Eleanor McCallie. $1.25
"The Banana Industry in the Philippines: An Informal Report," Frances Moore Lappé and Eleanor McCallie. $1.95

Food First Slideshow/Filmstrip in a visually positive and powerful portayal demonstrates that the cause of hunger is not scarcity but the increasing concentration of control over food-producing resources, 30 minutes. $89 (slideshow), $34 (filmstrip)

Write for our free publications catalogue.
All publications orders must be prepaid.

New from the authors of *Food First*

AID AS OBSTACLE

Twenty Questions about our Foreign Aid and the Hungry

by Frances Moore Lappé,
Joseph Collins,
David Kinley

Our most important contribution to ending hunger and poverty may be ending U.S. government aid to most third world governments.

Blue-ribbon commissions, representing some of the world's most powerful institutions, are calling for massive increases in official aid. Many concerned Americans will applaud such recommendations —until they read this myth-shattering book. Here, for the first time, Americans receive documented answers to such questions as:

- Doesn't our aid go to the countries with the greatest poverty?
- Hasn't our aid been reformed to target the needs of the poor?
- Aren't aid programs that focus on the small farmers a good way to end hunger?
- Doesn't our food aid feed the hungry?
- If not by official foreign aid, how can we help?

200 pages, with primer, charts, tables, photographs. $4.95. Frances Moore Lappé is author of *Diet for a Small Planet* (Ballantine, 1975). She and Joseph Collins (with Cary Fowler) are authors of *Food First: Beyond the Myth of Scarcity* (Ballantine, 1979) and are co-founders of the Institute for Food and Development Policy. David Kinley coordinates the Institute's Aid Education Project. The Institute is a not-for-profit research, documentation and education center focusing on food and agriculture.

Aid as Obstacle: Twenty Questions about our Foreign Aid and the Hungry
1-6 copies, $4.95; 7-49 copies, $3.95; 50 or more copies, $2.95.

____
name
____
address
____
city/state/zip

Make checks payable to
Institute for Food and Development Policy
2588 Mission Street
San Francisco, California 94110 U.S.A.
(415) 648-6090

ifdp

All orders must be prepaid. All foreign orders must be paid in U.S. funds with an International Money Order or drawn on a U.S. bank.

Subtotal ______
Add 6% of subtotal for California residents sales tax ______
Add 70¢ for special U.S. postage handling ______
Add 20% of subtotal for foreign surface mail ______
Add 40% of subtotal for foreign air mail ______
TOTAL ENCLOSED ______